1.1 Properties of Life

1

Characteristics of Life

In order to study something effectively, we need to be able to define it. In a biology class, learning about living things, it seems fitting to start out with the characteristics that make a living thing different from other non-living components of the universe.

How Do We Define "Life"?

We'll use the NASA guidelines for defining life, as listed below. We'll dive into each one of these guidelines in more detail in this section.

» Organization
» Genetic Information
» Reproduction
» Growth
» Homeostasis
» Regulation
» Response to Stimuli
» Processing Energy

Organization

All living things are made up of one or more cells, which is why biologists call cells the basic "unit" of life. Living organisms have a hierarchical structure, where smaller components work together to form larger, more complex structures with specialized functions. The molecules that make up a cell are arranged and built, or organized, into complex structures that facilitate specific functions, as you can see in Figure 1.1.

For example, the organization of atoms into a molecule of DNA is distinct from the organization of atoms into a molecule of fat. The difference in organization, or structure, gives these molecules different functions. Cells can then possess the characteristics of life because the structure of the molecules of which cells are composed are able to interact with each other to produce chemical reactions that make the processes of life possible.

Figure 1.1. The Organization of Living Things

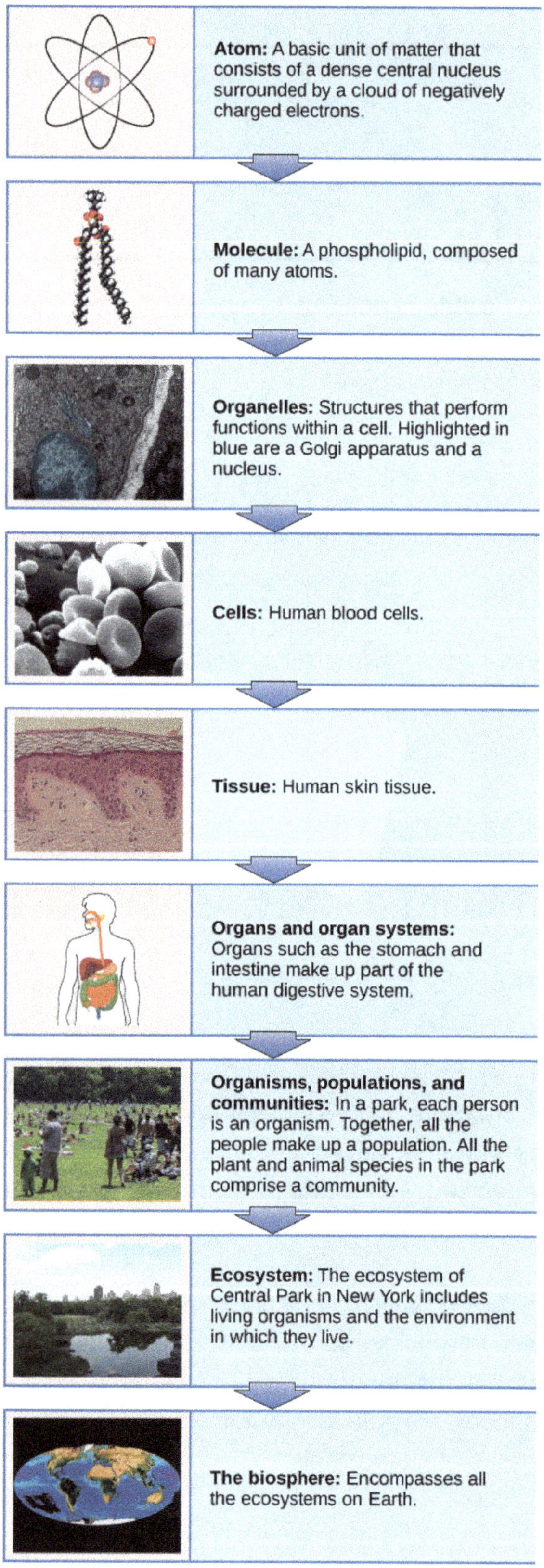

In a multicellular organism like a human, cells group together to form tissues such as muscle or nerve tissue. Each of these cells has its own unique organizational pattern that allows for its specific function. Tissues combine to form organs like the heart or brain, and organs work together in organ systems like the cardiovascular or nervous system.

What we've been skirting around so far is the idea of emergent properties where a new process or behavior is possible when components interact, or put another way, living things are more than the sum of their parts. Emergent properties are a result of the organization and interactions of the parts that make up the whole. Like how a cell is alive, but if we isolate the molecules that make up a cell, like DNA and proteins, those molecules, by themselves, are not alive.

You'll likely hear the phrase "form determines function" or "structure determines function" and this applies to many different levels of life, from the molecular structure to the structure of organs or parts of the body. What do you think would be the result if the structure of something was changed?

Genetic Information

All living things possess some form of genetic material that contains the instructions for how to build the molecules necessary for life to occur. You've likely heard of this molecule: DNA or deoxyribose nucleic acid. We'll spend time talking about the structure and function of this molecule later, but for now, we need to know that all living things contain DNA, and it functions as the genetic information molecule. DNA stores the instructions for how to carry

Cell Biology for Health Occupations

Active Learning Workbook

Third Edition

Holly Schiefelbein

Chemeketa Press | Salem, Oregon

Cell Biology for Health Occupations: Active Learning Workbook
Third Edition

ISBN-13: 978-1-955499-57-6

Chemeketa Press
Chemeketa Community College
4000 Lancaster Dr NE
Salem, Oregon 97305
collegepress@chemeketa.edu
chemeketapress.org

Cover design by Ronald Cox
Interior design by Ronald Cox and Abbey Gaterud

A full list of credits appears on pages 205–207 and constitutes an extension of this copyright page.

References to website URLs were accurate at the time of writing. Neither the author nor Chemeketa Press is responsible for URLs that have changed or expired since the manuscript was prepared.

Printed in the United States of America.

Land Acknowledgment

Chemeketa Press is located on the land of the Kalapuya, who today are represented by the Confederated Tribes of the Grand Ronde and the Confederated Tribes of the Siletz Indians, whose relationship with this land continues to this day. We offer gratitude for the land itself, for those who have stewarded it for generations, and for the opportunity to study, learn, work, and be in community on this land. We acknowledge that our College's history, like many others, is fundamentally tied to the first colonial developments in the Willamette Valley in Oregon. Finally, we respectfully acknowledge and honor past, present, and future Indigenous students of Chemeketa Community College.

Contents

out all of life's processes. Each one of your billions of cells has the exact same sequence of DNA, and unless you have an identical twin, your sequence of DNA is unique to you, as you can see in Figure 1.2.

Figure 1.2. Genetic Information in Living Things

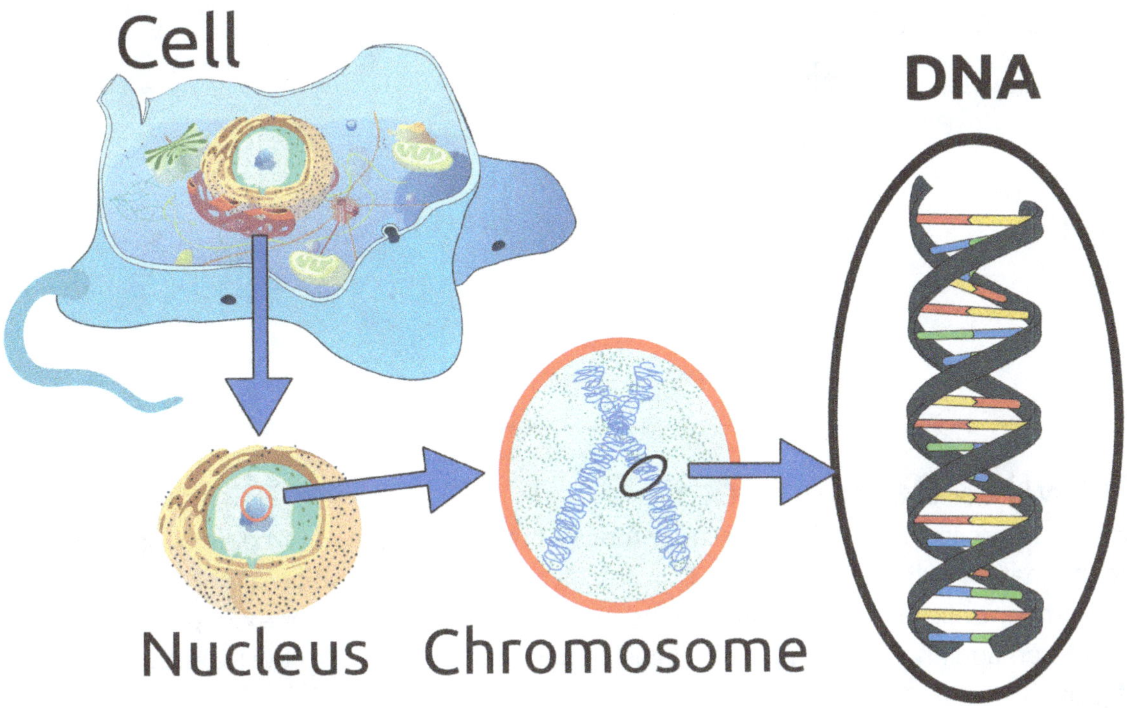

Reproduction

Living things make more of their own kind in the process of reproduction. Reproduction can be asexual, like a cell dividing into two identical copies of itself. Or, reproduction can be sexual, in which a new organism is produced which is a combination of DNA from two parents.

1. Describe the significance of the genetic material in living organisms and how it contributes to reproduction.

2. Viruses are essentially genetic information contained within a protein coat, as you can see in Figure 1.3. They can reproduce, but only once they are inside another cell after hijacking the cell's machinery. Are viruses alive?

Figure 1.3. The Components of a Virus. *Created with Biorender.*

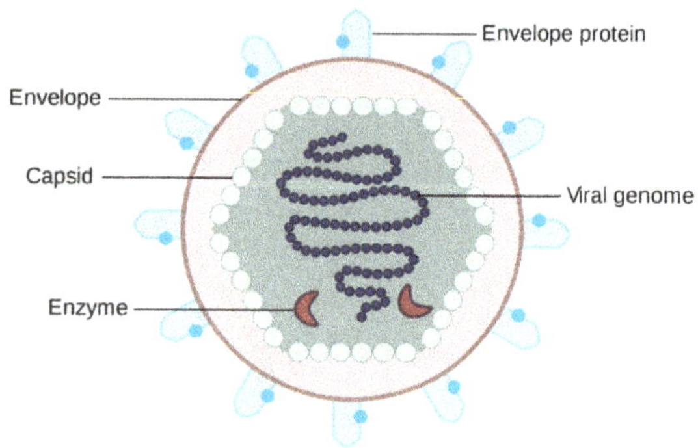

Growth & Development

Growth refers to the increase in size or mass of an organism over time. In simpler terms, it means that living things get bigger. This can involve the multiplication of cells, an increase in the number of cells, or an increase in the size of individual cells. For example, a baby growing into an adult human or a plant increasing in height and adding new leaves are instances of growth.

Development, on the other hand, is a broader term that encompasses not just the increase in size but also the changes in form, structure, and function that occur as an organism progresses through its life cycle. It involves the process of maturation and differentiation of cells and tissues. For instance, the transformation of a fertilized egg into a complex, multicellular organism with specialized organs and tissues is an example of development.

3. Compare and contrast "growth" and "development." Provide an example of each.

4. What role do you think DNA (genetic information) plays in the processes of growth and development?

Homeostasis, Regulation, and Response to Stimuli

Homeostasis

At the cellular level, homeostasis is a fundamental process essential for the proper functioning and survival of individual cells. Cells must regulate internal conditions, such as ion concentrations, pH levels, and nutrient availability, to maintain a stable and conducive environment for biochemical reactions. If these biochemical reactions don't occur at the right time, place, and rate, homeostasis cannot be maintained, and thus cells will stop living.

Regulation

Regulation is a fundamental property of life that encompasses the ability of living organisms to control and maintain their internal environment despite external changes. This dynamic process ensures stability (homeostasis) can be maintained within the organism despite external fluctuations and is crucial for various life processes.

At the molecular level, enzymes play a pivotal role in regulating the homeostasis of the cellular environment that allows for the chemical reaction essential for life to occur. Cells maintain homeostasis by tightly controlling the concentrations of ions, like Na^+, molecules, like glucose, and nutrients through selective membrane permeability and active transport mechanisms.

On a broader scale, organisms regulate physiological functions such as temperature, pH, and water balance to create an internal environment conducive to life. For example, the human body maintains a constant core temperature through mechanisms like sweating or shivering in response to external temperature changes. If we didn't keep this constant temperature, the chemical reactions occurring in our cells, keeping us alive, would not function properly, as you can see in Figure 1.4.

Figure 1.4. Homeostasis Through Negative Feedback

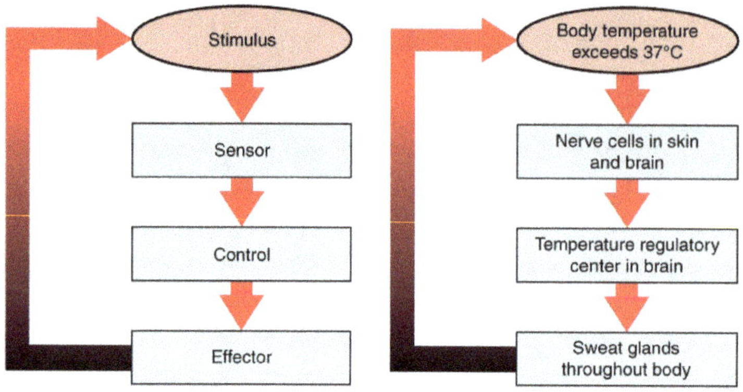

(a) Negative feedback loop (b) Body temperature regulation

It's not uncommon to get the concepts of "homeostasis" and "regulation" mixed up. Homeostasis comes from Greek roots, meaning "similar" (homeo) and "staying still" (stasis). So, homeostasis is the *state of being* that is held at a relatively constant level, often called a set point. Whereas regulation is the *processes* that are activated to maintain that constant state. For example, humans maintain a relatively constant body temperature of around 37 degrees Celsius. That temperature is the homeostasis state. Regulation would be the processes, like sweating or shivering, that maintain that constant body temperature.

5. Why do living things need to maintain homeostasis?

6. How are homeostasis and regulation related? How are they distinct, separate concepts?

Response to Stimuli

The maintenance of homeostasis through regulation wouldn't be possible if living things, including cells, weren't able to respond to both internal and external stimuli. In the context of biology, a stimulus refers to any signal or factor that elicits a response from an organism or cell. Stimuli can come in various forms, including changes in temperature, light, pressure, chemicals, or other environmental conditions. Essentially, a stimulus is something that triggers a reaction or a change in an organism's behavior or internal state. Stimuli will often trigger the processes of regulation, processes which enable the maintenance of homeostasis.

To demonstrate this concept, let's think about what happens when we hyperventilate. Hyperventilation is when we breathe faster than is required for our current body state, and it is often triggered by anxiety. When we hyperventilate, we get lightheaded. What's the best thing to do for someone when they're hyperventilating to prevent them from passing out? Give them a paper bag to breathe into. Let's investigate this phenomenon as it relates to homeostasis, regulation, and response to stimuli.

Normally, breathing helps regulate the levels of oxygen and carbon dioxide in the body. When you hyperventilate, you expel more CO_2 than your body is producing, leading to a deviation from the normal homeostasis range for CO_2 concentration. This deviation from homeostasis can result in several physiological changes (regulation kicks in to return blood CO_2 levels to the homeostasis range). The stimuli that starts this process is the change in CO_2 concentration of your blood, detected by your nervous system.

So why do we feel lightheaded? A decrease in CO_2 levels (stimuli) causes blood vessels to constrict (regulation). The constriction of blood vessels is intended to decrease the amount of CO_2 that is delivered to the lungs, thus bringing the concentration of CO_2 in the blood back to a normal range. However, this constriction of blood vessels decreases blood flow to the brain, which decreases the amount of oxygen being delivered to your brain. With less oxygen making it to the brain, we can start to feel lightheaded and dizzy, and if we don't return to a normal state, we can pass out. By breathing into a paper bag, we are re-inhaling the CO_2 we just exhaled, which increases the concentration of CO_2 in our blood, which then causes our blood vessels to dilate and returns blood flow to the brain, as you can see in Figure 1.5.

Figure 1.5. CO_2 Regulation Cycle

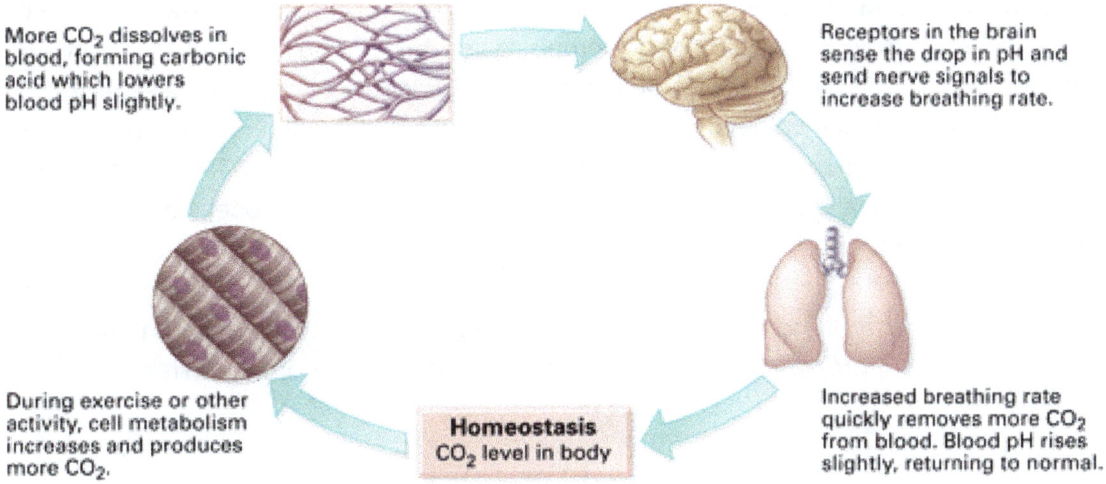

7. Response to stimuli includes responding to both internal and external stimuli. Why do living things need to monitor internal stimuli? Can you think of an example besides the one given about CO_2 levels?

8. Do all living things sense stimuli the same way? Give an example of an organism that you think senses stimuli in a different manner than humans.

9. Can homeostasis be maintained without the ability to sense both external and internal stimuli? Explain your thinking.

Processing Energy

The concept of "processing energy" in terms of the defining characteristics of life refers to the fundamental requirement for living organisms to acquire, convert, and utilize energy for various biological processes. Living things need energy to perform activities such as growth, reproduction, maintaining internal stability (homeostasis), responding to stimuli, and carrying out metabolic reactions. The acquisition of energy typically occurs through processes like photosynthesis in plants or the consumption of food in animals. Once acquired, energy is then converted and stored in a usable form, often as adenosine triphosphate (ATP), to power cellular activities and chemical reactions. The ability of living organisms to process energy is vital for sustaining life and all the functions associated with it.

10. Without energy, living organisms cannot maintain homeostasis. Do you think the emergent property of "life" would be possible without homeostasis?

11. How do animals and plants differ in the source of their energy? Do they both process energy?

12. What is an emergent property? Give an example of an emergent property that a cell has, that the molecules that make up a cell do not possess.

Multiple Choice Review

1. A scientist discovers a new organism in a deep-sea environment. Which of the following properties would they most likely investigate first to determine if it is a living organism?
   a. Its ability to conduct electricity
   b. Its response to environmental changes
   c. The presence of carbon in its structure
   d. Its color and shape

2. If a plant is placed near a window, it tends to grow towards the light. This behavior is an example of which characteristic of life?
   a. Homeostasis
   b. Metabolism
   c. Response to stimuli
   d. Reproduction

3. When a human body temperature rises due to an infection, the body sweats to cool down. This is an example of which life property?
   a. Cellular organization
   b. Energy utilization
   c. Homeostasis
   d. Growth and development
   e. Regulation

4. In an ecosystem, plants use sunlight to make food, animals eat plants or other animals, and decomposers break down dead matter. This scenario primarily illustrates which property of life?
   a. Energy processing
   b. Cellular organization
   c. Reproduction
   d. Regulation

5. A biologist observes that a certain species of bacteria divides every 20 minutes under optimal conditions. This rapid division primarily demonstrates the bacteria's ability to:
   a. Response to stimuli
   b. Maintain homeostasis
   c. Reproduce
   d. Process energy

1.2 The Scientific Method

What is the Scientific Method?

The scientific method (or scientific process) is simply a formalized mechanism for answering questions about the natural world through the collection of data. The data collected during the scientific process should be objective measurements of tangible natural phenomena. While there are many ways of knowing about the world, the main purpose of the scientific method is to produce quantifiable data that has been objectively measured to better understand the natural world.

1. What are some questions that would be outside the scope of the scientific method?

2. What does "quantifiable" mean? Why would this be an important aspect of gathering information about the world?

Let's break down the components of the scientific method. While this process is often shown as a cycle, there are many ways to move through the process, and backtracking, rethinking, or reevaluating in light of new evidence and data is common, as you can see in Figure 1.6.

Figure 1.6. The Scientific Method

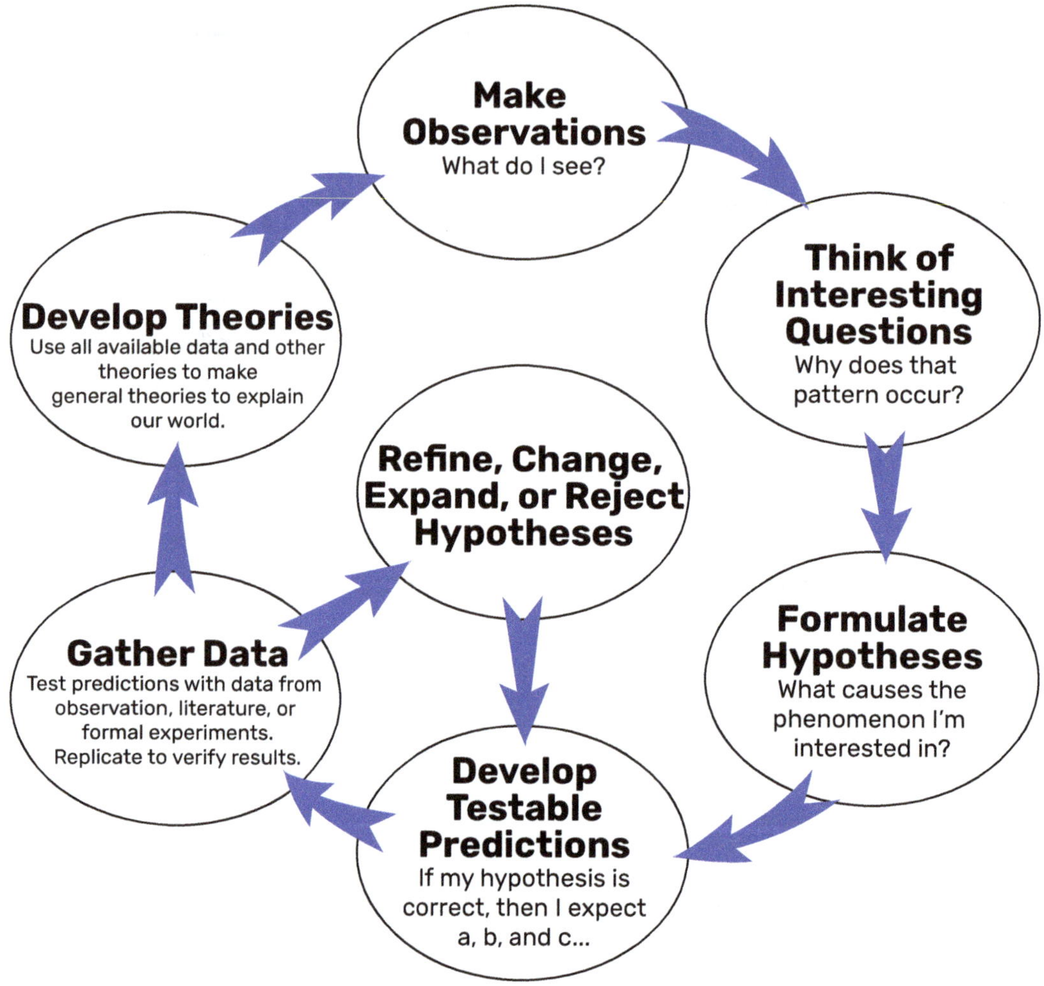

We're spending time here learning about the scientific method because all of what we will learn in science classes is the result of someone moving through the scientific process and sharing their findings. Additionally, as future healthcare professionals practicing evidence-based medicine, it will be important to evaluate claims made about treatments or to critically evaluate the research studies to assess the validity of the results.

- **Observation/Question:** Identify a phenomenon or ask a question about the natural world.
  - *Example*: You observe that patients in a hospital's intensive care unit (ICU) who receive a certain type of therapy seem to recover more quickly from surgery.
- **Hypothesis:** Formulate a testable statement explaining the mechanisms causing the phenomenon and predicting the outcome of the experiment.
  - *Example*: Based on your research, you hypothesize that the specific therapy given in the ICU is contributing to the quicker recovery of patients after surgery by increasing blood flow to the surgery site.

» **Experiment:** Design and conduct a controlled investigation to gather data to test the hypothesis.
   » *Example*: Design a controlled study comparing the recovery rates of patients who receive the standard therapy and those who receive the specific therapy in question.

Let's take some time to explore the components of an experiment. It's valuable to understand these components to interpret the validity of scientific findings. Experiments are designed to test the effects of the **independent variable** on the **dependent variable**. The independent variable is the variable that the researcher intentionally manipulates or changes to observe its effect on the dependent variable. The dependent variable is the variable that is measured or observed to determine the effects of the independent variable. It is the outcome or response that researchers are interested in studying. In the ICU example, the independent variable is the new therapy, and the dependent variable is the recovery time for patients after surgery. Experiments will utilize control groups and one or more treatment groups. Control groups and treatment groups are essential in experimental design to help researchers draw meaningful conclusions about the effects of an independent variable.

A **control group** in an experiment is exposed to the same experimental conditions as the treatment group but does not receive the experimental treatment (independent variable). The control group provides a baseline against which the effects of the independent variable can be compared. By keeping all conditions the same except for the independent variable, researchers can better attribute any observed effects to the treatment. In healthcare studies, the control group is often called the placebo group. A **placebo** is a substance or treatment that has no therapeutic effect but is given to a patient as if it were a real treatment. Placebos help researchers separate the psychological or subjective effects of treatment from its specific pharmacological effects. This is important in understanding how much of a treatment's effectiveness is due to the actual substance and how much is influenced by psychological factors like expectation and belief. In the ICU therapy study, the control group might consist of patients who receive the standard post-surgery recovery treatment commonly used in the hospital but not the specific therapy being investigated.

The **treatment group** in an experiment is exposed to the experimental treatment of the independent variable. The treatment group allows researchers to observe and measure the effects of the independent variable on the dependent variable. Any differences between the treatment group and the control group can be attributed to the specific treatment being tested. In the ICU therapy study, the treatment group would include patients who receive the specific post-surgery recovery therapy that the researcher is investigating. This group's outcomes are compared to those of the control group to assess the impact of the therapy.

It's also important to maintain consistency between the treatment group and the control groups as much as possible. This is done to minimize any impact of other factors besides the independent variable on the dependent variable. The aspects that are held constant between the treatment and the control group are called controlled variables. **Controlled variables** are factors in an experiment that are intentionally kept constant or controlled to prevent them from influencing the results. In the ICU therapy study, the type of surgery performed on patients (e.g., a specific procedure—appendectomy versus mastectomy) could be a controlled variable. By keeping this constant, you ensure that any differences in recovery rates are more likely to be attributed to the independent variable (the specific therapy) rather than variations in surgical procedures.

- **Communication:** Share findings through publications, presentations, or other means to contribute to scientific knowledge.
  - *Example*: Share your findings with the medical community through research papers or presentations. This step allows other healthcare professionals to assess the potential benefits of the therapy and consider its implementation in other settings.

- **Further Inquiry:** Propose additional research questions or experiments to deepen understanding or address related issues.
  - *Example*: Other researchers may be inspired to delve deeper into the mechanism of action of the therapy, its long-term effects, or its applicability to different patient populations.

3. Consider a study where researchers want to determine if drinking green tea affects alertness in students. They hypothesize that students who drink green tea will show increased alertness compared to those who don't. To test this, they divide students into two groups: one group drinks green tea, and the other drinks a placebo beverage without caffeine. Both groups then undergo tests to measure their alertness levels.
   a. What hypothesis is being tested here?
   b. What is the dependent variable in this experiment?
   c. What is the independent variable in this experiment?
   d. What is the treatment group?
   e. What is the control group?

4. Design a study to explore the impact of sleep duration on white blood cell count, using the scientific method. Describe how you would develop a hypothesis, plan the experimental design, specify the methods for data collection and analysis, and determine how to interpret the results. Also, discuss how you would handle ethical issues related to the study, such as participant consent and confidentiality.

Vaccines

Now that we have an understanding of the components of the scientific method, let's walk through a real-world example of science not being conducted ethically or objectively and the ripple effects this has had on our society and health care system. We're going to explore the vaccine and autism debate.

Overview of vaccines

A vaccine is a substance that, when administered, provides immunity to a specific disease-causing agent, like a bacteria, virus, or other pathogen. Vaccines typically contain a small, harmless part of the disease-causing virus or bacteria. When a new pathogen or disease enters the body, it introduces a new antigen. For every new antigen, the body needs to build a specific antibody that can attach to the antigen and defeat the pathogen (Figure 1.7a). When a person is vaccinated, their immune system is exposed to this harmless part of the bacteria or virus, stimulating the production of new antibodies and memory immune cells. If the person is later exposed to the actual disease-causing microorganism, their immune system can recognize it and mount a faster and more effective defense, preventing or mitigating the disease (Figure 1.7b).

Figure 1.7a and 1.7b. Vaccine Function in the Body

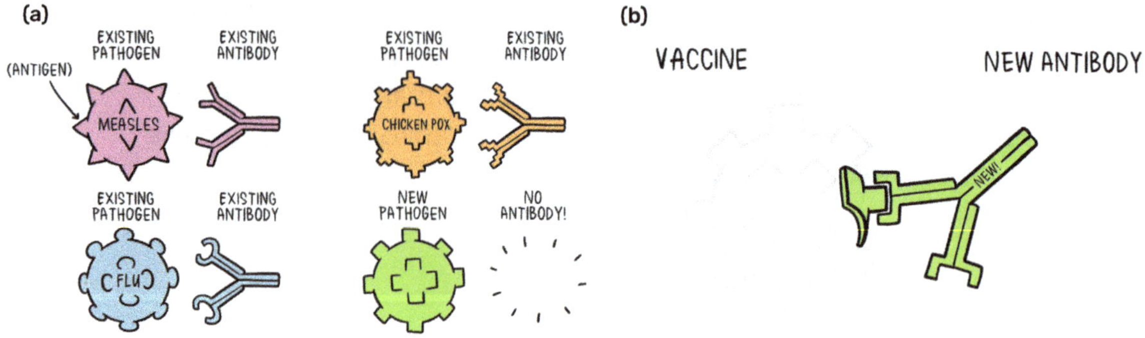

There is evidence of a type of vaccination—called variolation—dating back to the eleventh century in India and China. In this practice, scabs from smallpox would be ground up and inhaled by healthy people, thus exposing them to the virus, inducing a mild disease, and preventing life-threatening disease if they were to come into contact with the smallpox virus at a later time.

Childhood vaccines have been credited with saving the lives of 4 million people every year. It's estimated around 50 million deaths can be prevented with vaccines between 2021 and 2030. There are over 25 safe and effective vaccines for a wide variety of diseases available.

Overview of ASD

Autism Spectrum Disorder (ASD) is an intricate neurodevelopmental condition marked by difficulties in social interaction, communication, and repetitive behaviors. Referred to as a "spectrum disorder," it encompasses many variations in the type and severity of symptoms individuals may exhibit.

Several studies have indicated an increase in the diagnosis of children with ASD over the last 20 years, while the diagnosis of other intellectual disabilities has remained relatively constant (Figure 1.8). In 1971, the combined Measles, Mumps, and Rubella vaccine was developed to provide three vaccines in one shot. By decreasing the number of doctor visits necessary to achieve immunity, the combined vaccine is intended to make it easier for more people to be vaccinated against three dangerous diseases in one visit.

Figure 1.8. Prevalence of Five Developmental Disabilities, Metropolitan Atlanta Developmental Disabilities Surveillance Program, 1991–2010

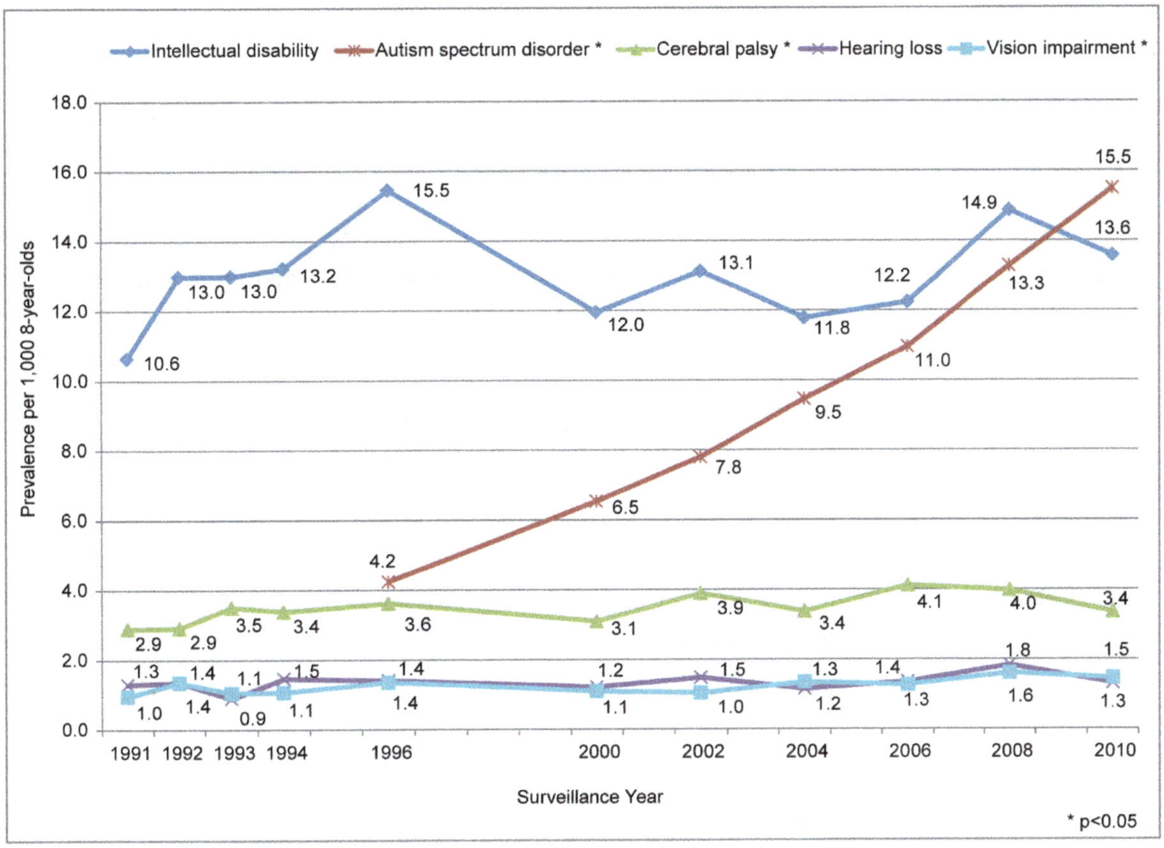

The Vaccines & Autism Conundrum

In 1998, Andrew Wakefield, a British surgeon and medical researcher, published a study in the prestigious medical journal *The Lancet*, which claimed to find a link between the measles, mumps, and rubella (MMR) vaccine and the development of autism in children. Wakefield's study suggested that the vaccine could cause a series of events leading to intestinal problems, which in turn resulted in autism. This publication sparked a major health scare, leading to a significant drop in vaccination rates worldwide.

However, investigations into Wakefield's research revealed numerous problems, including a small sample size, lack of a control group, and speculative conclusions that were not supported by the data. Furthermore, it was discovered that Wakefield had multiple undeclared conflicts of interest and had manipulated evidence. Several independent researchers attempted to replicate Wakefield's findings but to no avail. *The Lancet* eventually retracted Wakefield's paper in 2010. To retract a journal article means to officially take back or withdraw the article from the scientific community after it has been published. It's like saying, "We made a mistake in publishing this, and we don't stand by it anymore."

1

5. What was the main claim of Andrew Wakefield's 1998 study published in *The Lancet*, and what impact did it have on public health perceptions regarding the MMR vaccine?

6. Why did the scientific community find issues with Wakefield's research, and what were the key problems identified in his study?

7. Explain the significance of *The Lancet* retracting Wakefield's paper in 2010 and what it implies about the validity of the research.

8. How do attempts by independent researchers to replicate Wakefield's findings contribute to the understanding of the scientific method's role in validating or disproving research results?

9. Do you think the general public still believes there to be a connection between vaccination and autism? Why do you think this is?

Multiple Choice Review

1. When conducting an experiment to test the effects of sunlight on plant growth, what would be considered the independent variable?
   a. Type of plant
   b. Amount of water given
   c. Amount of sunlight
   d. Growth of the plant

2. If you wanted to test the hypothesis that temperature affects the rate of a chemical reaction, what would be a suitable control?
   a. A reaction at room temperature
   b. A reaction using a different chemical
   c. A reaction at the highest temperature
   d. A reaction without any reactants

3. In an experiment to determine if studying with music improves test scores, what would be the dependent variable?
   a. Type of music played
   b. Number of students in the study
   c. Test scores of the students
   d. Amount of time spent studying

4. Which of the following best describes the role of a hypothesis in a scientific experiment?
   a. It is a detailed record of the experiment's results.
   b. It is the process used to measure the outcome of the experiment.
   c. It is a prediction or explanation that can be tested through study and experimentation.
   d. It is a summary of the research and findings of the experiment.

5. When analyzing data from an experiment to test a new drug's effectiveness, which of the following steps should come first?
   a. Publication of the results
   b. Development of a hypothesis
   c. Statistical analysis of the data
   d. Formation of a conclusion based on the data

1

6. In a scientific study to explore the relationship between sleep and memory, if researchers manipulate the hours of sleep participants get, what kind of variable is sleep duration?
   a. Dependent variable
   b. Independent variable
   c. Controlled variable
   d. Extraneous variable

7. If a scientist wants to ensure that an experiment testing a new fertilizer's effect on plant growth is valid, which of the following should they avoid?
   a. Using plants of the same species
   b. Changing the amount of water each plant receives
   c. Keeping the amount of sunlight constant
   d. Measuring the growth of the plants in centimeters

8. Why is it important to have a control group in a scientific experiment?
   a. To ensure that the experiment can be repeated
   b. To provide a standard against which the experimental results can be compared
   c. To increase the number of participants in the experiment
   d. To prove the experimental hypothesis is correct

9. When designing an experiment to test how exercise affects heart rate, what would be an appropriate control activity?
   a. Running at a moderate pace
   b. Sitting quietly
   c. Drinking caffeinated beverages
   d. Sleeping

10. After conducting an experiment, you find that the results do not support your hypothesis. What is the most appropriate next step?
    a. Discard the entire experiment and start a new topic
    b. Modify the hypothesis based on the results and conduct another experiment
    c. Ignore the results and report the expected outcome
    d. Publish the results without peer review

1.3 Atomic Structure

Matter is anything that has mass and takes up space. This includes living and nonliving things. The air around you has mass and takes up space, but it's not alive. All matter is composed of atoms of particular elements; understanding the structure and properties of atoms will help us understand the molecules commonly found in living organisms. Atoms are the smallest units or pieces of matter that retain the characteristics of their element. Atoms are often called building blocks since atoms in living things are often combined together into molecules by chemical bonds.

Subatomic Particles

To understand how atoms interact with each other, we need to first understand what makes up an atom. Atoms are composed of subatomic particles: neutrons, protons, and electrons. Each of these subatomic particles plays an important role in the function and properties of an atom, including how it will behave and interact with other atoms.

The two general regions of an atom are the nucleus and the electron cloud (sometimes called the electron orbitals). The nucleus is the center of the atom, this is what contains the particles which have significant mass: neutrons and protons. You can see the atomic structure in Figure 1.9.

Figure 1.9. The Structure of an Atom

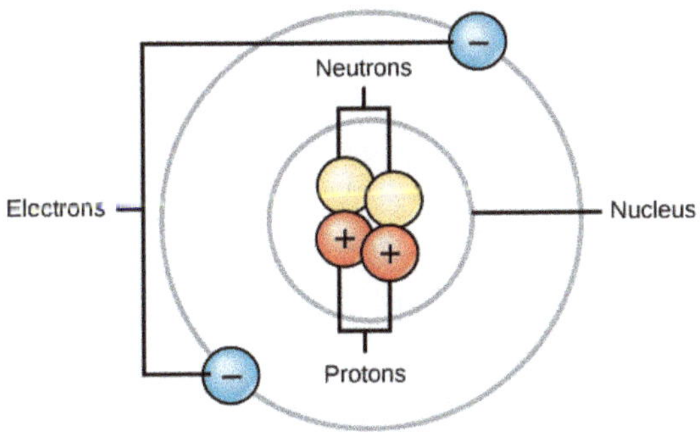

Protons

The number of protons defines which element we are working with. If the number of protons in the nucleus were to change, then we'd be changing the element we're working with. Protons have a mass of 1 AMU (atomic mass unit) and have a charge of +1. The number of protons in an atom is labeled as the atomic number on the periodic table, as seen in Figure 1.10.

Figure 1.10. Periodic Table Element Overview

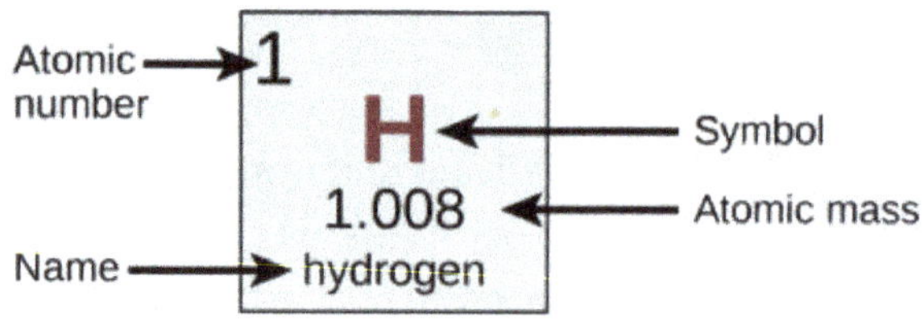

Neutrons

Neutrons are also found in the nucleus, but they are neutral, meaning they do not have a charge or have a charge of zero. Neutrons act as "glue" in the nucleus, holding everything together. They have a mass of 1 AMU. Changing the number of neutrons in an atom is how we get **isotopes**. The number of neutrons can change, and as long as the number of protons remains the same, we get atoms of the same element but with different atomic masses.

Atomic Mass and Isotopes

The atomic mass of an element is the sum of the number of protons and neutrons, as seen in Figure 1.11. For example, most atoms of Carbon have 6 protons and 6 neutrons, so the atomic mass of these atoms would be 12 AMU. But, there are a small percentage of Carbon atoms that have 7 neutrons, which would make their atomic mass 13 AMU. These are isotopes of Carbon, since the number of protons has remained constant, but the number of neutrons is no longer equal to the number of protons.

Figure 1.11. Carbon Isotopes Comparison

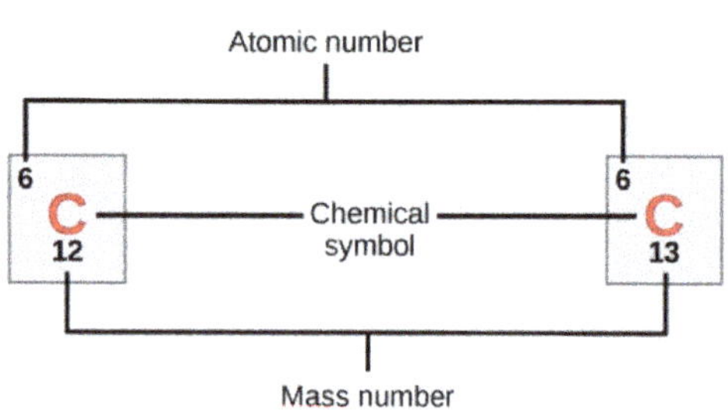

Electrons

Electrons are subatomic particles that are, at least for our purposes, massless. We can assume they have a mass of 0 AMU. Electrons have a charge of -1, however. This means they influence the charge of an atom. The charge of an atom can be calculated by subtracting the number of electrons from the number of protons.

Atomic Charge = number of protons - number of electrons

In our class, *unless it's specifically indicated otherwise*, we can assume we're working with an atom that has a charge of 0 when answering questions and drawing atomic models. If an atom doesn't have an equal number of protons and electrons, that atom is an **ion**. When there are more electrons than protons, the charge will be negative. With more protons than electrons, the charge will be positive.

Figure 1.12. Periodic Table Electron Configurations

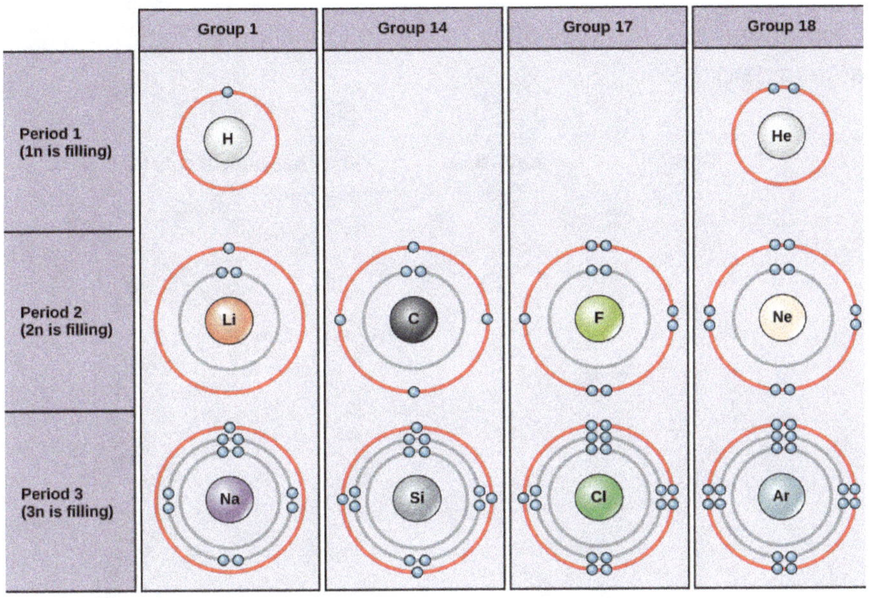

Electrons orbit around the nucleus in what we call shells or orbitals. We use this terminology to demonstrate a concept when, in reality, the behavior of electrons is much more complicated than we need to get into. If it helps, think of the electron shells like an onion, and the nucleus of the atom is at the center of the onion (Figure 1.12). The first shell can only hold two electrons since it's the smallest. If an atom has more than two electrons, it starts filling up the second shell, which can hold eight electrons. Once the second shell is full, electrons will start filling up the third shell, which can also hold eight electrons. Most elements that are important to life typically don't have electron shells past the third shell.

Electrons are important for determining the chemical reactivity of an atom, or how and what type of bonds it makes, which we'll go into in Lab 2.

1

1. How many protons, neutrons, and electrons are contained in an atom of Argon, which has an atomic number of 18 and a mass number of 40?

2. In Table 1.1 use the information in the previous reading to identify the properties of subatomic particles.

Table 1.1.

Subatomic Particle	**Mass**	**Charge**	**Location**	**What happens if this is changed?**

3. The smallest unit of an element that still retains the distinctive properties of that element is a(n)...

4. The subatomic particle that gives an element its distinctive properties is its number of...

5. Draw a Bohr model of Carbon below. Figure 1.9 shows a Bohr model.

6. What happens when an atom has a different number of protons and electrons?

7. What is the importance of electrons?

Multiple Choice Review

1. Which subatomic particle determines the chemical properties of an element?
   a. Neutron
   b. Electron
   c. Proton
   d. Quark

2. If an atom has 17 protons and 18 neutrons, what is its atomic number?
   a. 17
   b. 18
   c. 35
   d. 36

1

3. An ion has 12 protons, 10 electrons, and 12 neutrons. What is the charge of the ion?
   a. +2
   b. -2
   c. +1
   d. -1

4. Which of the following isotopes has the same number of neutrons as carbon-14?
   a. Carbon-12
   b. Nitrogen-14
   c. Oxygen-16
   d. Nitrogen-15

5. An element has an atomic number of 15. How many valence electrons does it have?
   a. 3
   b. 5
   c. 15
   d. 7

6. What is the mass number of an atom that contains 20 protons, 20 neutrons, and 20 electrons?
   a. 20
   b. 40
   c. 60
   d. 80

7. An atom with a mass number of 23 and 11 protons must have how many neutrons?
   a. 11
   b. 12
   c. 23
   d. 34

8. If an atom loses an electron, what happens to its overall charge?
   a. It becomes positively charged.
   b. It becomes negatively charged.
   c. It remains neutral.
   d. It becomes an isotope.

1.4 States of Matter

All matter can exist in one of three phases: solid, liquid, or gas (Figure 1.13). These phases are determined by the arrangement of the atoms or molecules within the material. Understanding the atomic structure of matter is crucial for understanding the behavior of materials in different phases.

In the **solid** phase, atoms or molecules are tightly packed and held in a fixed position by intermolecular forces. This results in a rigid structure with a definite shape and volume. The intermolecular forces that hold solids together are typically the strongest of the three phases, due to the close proximity of the atoms or molecules. These forces can be broken down into two types: covalent and ionic bonds. Covalent bonds occur when two atoms share electrons, resulting in a strong bond. Ionic bonds occur when atoms transfer electrons to form ions, which are then held together by electrostatic forces. We will investigate how these types of bonds form later.

In the **liquid** phase, atoms or molecules are more loosely packed than in a solid but are still held together by intermolecular forces. These forces are weaker than in the solid phase, which allows molecules to move past one another, resulting in a fluid or liquid-like behavior. Liquids take on the shape of their container but still maintain a defined volume.

In the **gas** phase, atoms or molecules are far apart and are not held together by intermolecular forces, resulting in a highly disordered and random structure. Gasses do not have a fixed shape or volume and can fill any container that they are placed in. Examples of gasses include the atmosphere, helium gas in balloons, and carbon dioxide when at room temperature (22 degrees Celsius).

Figure 1.13. States of Matter

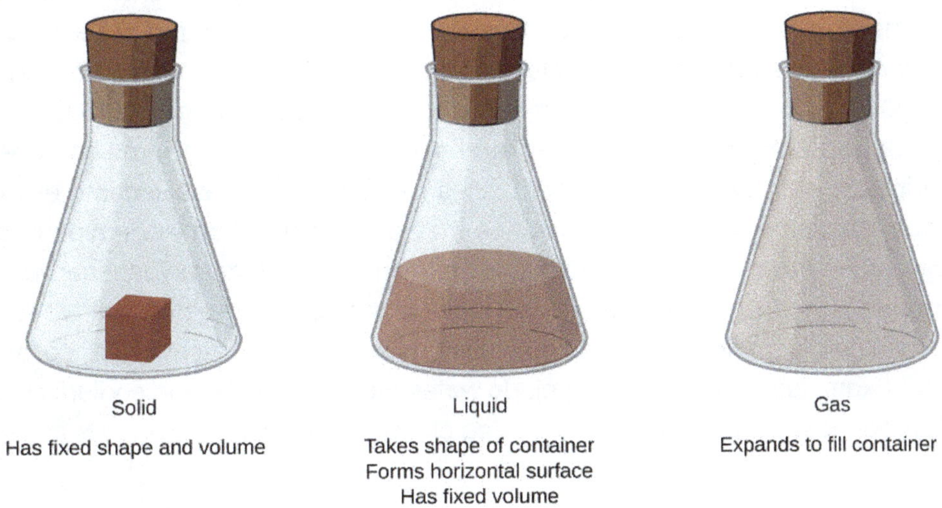

1. Do you think you'll come into contact with the three different states of matter in a health care setting? Provide an example of how you may encounter these three different states of matter in your healthcare career.

Classifications of Matter

Learning about the classifications of matter is important when studying biology because all living organisms are made up of matter. Understanding the structure and properties of matter is essential for understanding the structure and properties of living things, as well as how they interact with each other and their environment.

Additionally, the human body is composed of different types of matter, including proteins, carbohydrates, lipids, and a variety of mixtures and solutions. Each of these types of matter has a unique structure and function, and understanding their properties is essential for understanding how they contribute to the overall function of cells and, thus, the body.

Pure Substances

Matter is anything that has mass and takes up space. We can classify matter into various categories based on its composition and properties. One way to classify matter is by dividing it into pure substances and mixtures. A **pure substance** is a type of matter that has a fixed composition and cannot be separated into simpler substances by physical means. For instance, glucose ($C_6H_{12}O_6$) is a pure substance since it has a constant composition of 25% Carbon atoms, 50% Hydrogen atoms, and 25% Oxygen atoms. All glucose has the same composition of Carbon, Hydrogen, and Oxygen. Pure substances can be further classified into elements and compounds. An element is a pure substance that is made up of only one type of atom and cannot be broken down *chemically*. Examples of elements include oxygen, carbon, and gold. A **compound** is a pure substance that is made up of two or more different types of atoms that are chemically combined in a fixed ratio. Glucose would be a pure substance, as well as a compound. Examples of compounds include water, carbon dioxide, and sodium chloride (table salt). Compounds can be broken down using chemical means.

1

2. What is the difference between a compound and an element?

3. Do you think there are many compounds found in the human body? Why or why not?

Mixtures

If a substance doesn't have a constant composition and is made up of multiple substances, it's called a mixture. A **mixture** is a type of matter that contains two or more substances that can be separated by *physical* means. Mixtures can be further classified into homogeneous and heterogeneous mixtures, as seen in Figure 1.14.

Figure 1.14. Matter Classification Flow Chart

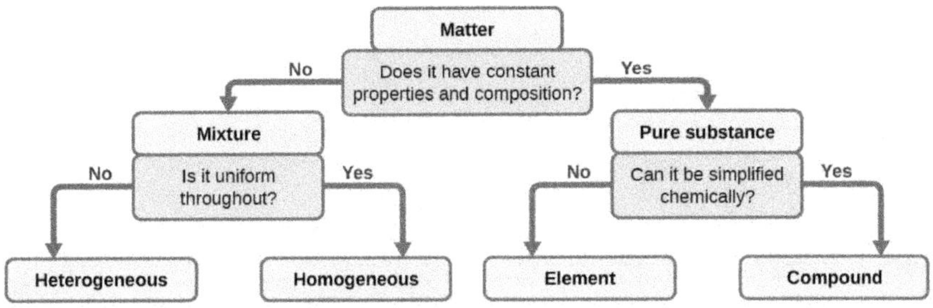

1

Homogenous Mixtures

Mixtures can be further classified into homogeneous and heterogeneous mixtures, as seen in Figure 1.15. A **homogeneous mixture** is a mixture in which the composition is uniform throughout. This means that the different substances that make up the mixture are evenly distributed and cannot be seen with the naked eye. Examples of homogeneous mixtures include saltwater and the atmosphere.

Solutions are a special type of homogeneous mixture in which one substance (the **solute**) is dissolved in another substance (the **solvent**). Solutions are made up of particles that are evenly distributed throughout the solvent. Solutions can be solid, liquid, or gas. An example of a solid solution is brass, which is a mixture of copper and zinc. An example of a liquid solution is sugar dissolved in water. An example of a gas solution is the atmosphere, which is a mixture of different gases, like N_2, CO_2, and O_2.

Figure 1.15. Types of Molecular Mixes

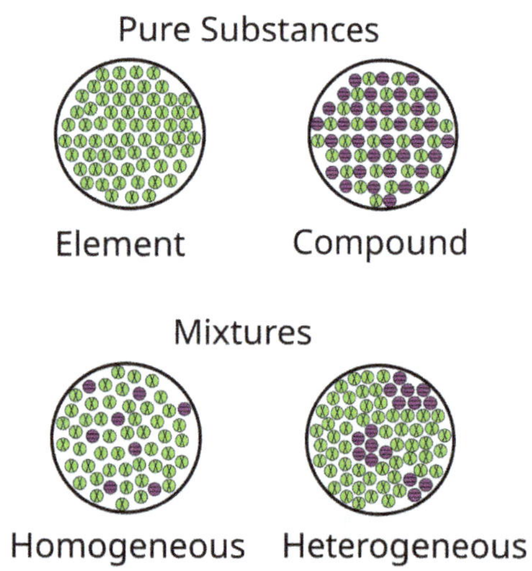

4. What distinguishes a mixture from a pure substance, and how can mixtures be physically separated into their individual components?

5. How does a homogeneous mixture differ from a heterogeneous mixture, and can you provide examples of each?

6. Explain what solutions are and describe how they fit into the classification of homogeneous mixtures, including examples in different states of matter.

Heterogenous Mixtures

A **heterogeneous mixture** is a mixture in which the composition is not uniform throughout. This means that the different substances that make up the mixture can be seen with the naked eye. A **suspension** is a type of heterogeneous mixture, meaning that its components are not evenly distributed throughout the mixture. In a suspension, small particles of one substance are suspended within another substance, typically a liquid. The particles in a suspension are typically visible to the naked eye and *will settle out over time if left undisturbed.* An example of a suspension is muddy water, where small particles of dirt or sediment are suspended in the water while it is in motion but will settle out if the water is left undisturbed.

A **colloid** is another type of heterogeneous mixture. In a colloid, small particles of one substance are dispersed throughout another substance, typically a liquid, but the particles are small enough that they do not settle out over time or are not visible to the naked eye. A colloid appears homogeneous because the particles are evenly distributed throughout the mixture, but it is actually a type of heterogeneous mixture. Examples of colloids include milk, fog, and paint.

1

Understanding the types of mixtures is important in preparing for a healthcare career for several reasons:

» **Medications:** Healthcare professionals often administer medications to patients. Many medications are mixtures of different compounds, and understanding the nature of these mixtures is essential to properly administer them.
» **Diagnostic tests & medical procedures:** Many diagnostic tests used in healthcare involve mixtures, such as blood samples. Understanding the properties of these mixtures is important in obtaining accurate test results.

7. Ringer's solution is administered via IV to replenish fluids and electrolytes in patients. It's composed of sodium chloride (NaCl), potassium chloride (KCl), calcium chloride ($CaCl_2$), and sodium lactate dissolved in water. What type of solution is Ringer's solution?

8. Blood consists of distinct components like red blood cells, white blood cells, platelets, and plasma, each with different properties and functions. These components are not uniformly distributed and can be separated through processes like centrifugation (spinning a vial of blood quickly). What type of mixture or solution is blood when it is in a vial? What about when it is in the body?

Multiple Choice Review

1. Which state of matter has a definite volume but no definite shape?
   a. Solid
   b. Liquid
   c. Gas
   d. Plasma

2. When salt is dissolved in water to form a solution, salt is the:
   a. Solvent
   b. Solute
   c. Suspension
   d. Colloid

3. In the context of solutions, what role does water play when salt is dissolved in it?
   a. Solute
   b. Solvent
   c. Precipitate
   d. Catalyst

4. What happens to the particles of a substance as it changes from a liquid to a gas?
   a. They move closer together.
   b. They move farther apart.
   c. They maintain their distance.
   d. They disappear.

5. What is the main difference between a colloid and a solution?
   a. The size of particles dispersed in the medium
   b. The electrical charge of particles
   c. The color of the resulting mixture
   d. The temperature at which they are formed

6. How would you classify a mixture of oil and water?
   a. Homogeneous mixture
   b. Heterogeneous mixture
   c. Compound
   d. Solution

2.1 The Periodic Table

The periodic table is a useful tool for understanding trends in the properties of elements. These trends can be explained by the atomic structure of each element, specifically the number of electrons in each shell and the number of valence electrons. Learning these trends can save you time in drawing out Bohr models to determine the number of electrons in the valence shell of an atom, and thus you'll be able to predict the number of bonds an atom can make much more quickly. You're not expected to memorize the names of each group (for example, you won't be tested on the name of elements in group 7), but it will be helpful to know that elements in group 7 have seven valence electrons, for example.

Electron Shells

The periodic table is organized by electron shells, with the elements in the same row (called periods, running horizontally) having the same number of electron shells. The electron shells, or energy levels, are represented by the rows of the periodic table. The first shell, closest to the nucleus, can hold up to 2 electrons, the second shell can hold up to 8 electrons, and the third shell can hold up to 8 electrons, as you can see in Figure 2.1.

Figure 2.1. Bohr Model of Oxygen

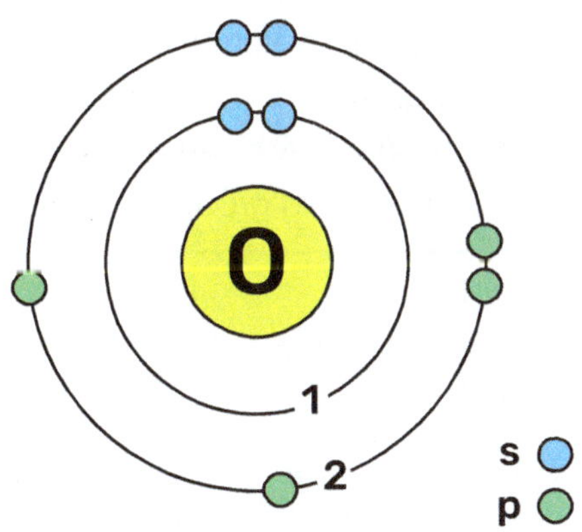

Valence Electrons

Valence electrons are the electrons in the outermost shell of an atom, also called the valence shell. They are important because they determine how an atom will react with other atoms, meaning what type and how many bonds an atom will make. Elements in the same group (groups are vertical columns) of the periodic table have the same number of valence electrons, which is why they often have similar chemical properties. Figure 2.2 shows an abbreviated periodic table of the common elements used in organic chemistry.

Figure 2.2. Abbreviated Periodic Table of Common Elements in Organic Molecules

Periodic table

of the most important elements in organic chemistry

Symbol	Density	Key
red: gas black: solid blue: liquid	red: density in $kg*m^{-3}$ black: density in $kg*dm^{-3}$	atomic number — standard atomic weight symbol name electronegativity — density

Period	I (1) Hydrogen & alkali metals	II (2) Alkaline earth metals	III (13) Triels	IV (14) Tetrels	V (15) Pnictogens	VI (16) Chalcogenes	VII (17) Halogens	VIII (18) Noble gases
1	1 1.0079 H Hydrogen 2.2 0.09							2 4.0026 He Helium --- 0.18
2	3 6.9675 Li Lithium 0.98 0.53	4 9.0122 Be Berylium 1.57 1.85	5 10.831 B Boron 2.04 2.46	6 12.011 C Carbon 2.55 2.26	7 14.007 N Nitrogen 3.04 1.25	8 15.999 O Oxygen 3.44 1.43	9 18.998 F Fluorine 3.98 1.7	10 6.9675 Ne Neon --- 0.53
3	11 22.990 Na Sodium 0.93 0.97	12 24.305 Mg Magnesium 1.31 1.74	13 26.981 Al Aluminum 1.61 2.70	14 28.085 Si Silicon 1.9 2.34	15 30.974 P Phosphorus 2.19 2.69	16 32.067 S Sulfur 2.58 2.07	17 35.451 Cl Chlorine 3.16 3.21	18 39.948 Ar Argon --- 1.78
4	19 39.098 K Potassium 0.82 0.86	20 40.078 Ca Calcium 1.0 1.55					35 79.904 Br Bromine 2.96 3.12	
5							53 126.904 I Iodine 2.1 4.94	

Elements in group 1 have 1 valence electron and will tend to give up that valence electron to form a +1 ion. Similarly, in group 2, these atoms will tend to give up their two valence electrons to form ions of +2. If we move over to elements in group 17 or (VII), these elements have seven valence electrons, so they tend to accept electrons from elements in groups 1 and 2.

1. What determines if an electron is a valence electron?

2. You may have heard of the octet rule. Explain what this is and why it's important.

Electronegativity

Electronegativity is the measure of an atom's ability to attract electrons toward itself in a chemical bond. Atoms with higher electronegativity values will attract electrons more strongly, resulting in a polar covalent bond or an ionic bond. Electronegativity values increase from left to right across a period and decrease from top to bottom down a group. We'll dive into how to use electronegativity to predict what type of bond will occur between two atoms later.

Understanding the trends in the periodic table can help us predict how elements will react and form chemical bonds with other elements. This is important in fields like biology and medicine, as it can help us understand the properties and behaviors of molecules and compounds that play important roles in biological systems. For example, we will talk about how the function of an important molecule in cells, called phospholipids, is possible because it has both polar and nonpolar bonds.

3. Explain the idea of electronegativity as if you were tutoring your middle school cousin.

4. Why do we need to know about electronegativity in this class?

2.2 Chemical Bonds

Atoms can combine with each other to form chemical bonds, which are interactions between the outermost electrons of different atoms. Atoms will make and break bonds during chemical reactions, which we'll talk more about later. When two or more atoms bond together, the result is a **molecule**, like a molecule of water which has an oxygen atom bonded to two hydrogen atoms (H_2O).

Types of Chemical Bonds

In some cases, electrons are transferred between atoms to form ionic bonds. In other cases, atoms share electrons to form covalent bonds. Understanding these types of chemical bonds is important for understanding the properties and behavior of molecules in biological systems. If we know what types of bonds are in a molecule, we can make some predictions about how that molecule will behave in a cell, and what its function is. Figure 2.3 shows how the type of bond that will occur between two atoms can be determined by calculating the difference in their electronegativity.

Figure 2.3. Electronegativity and Bond Types

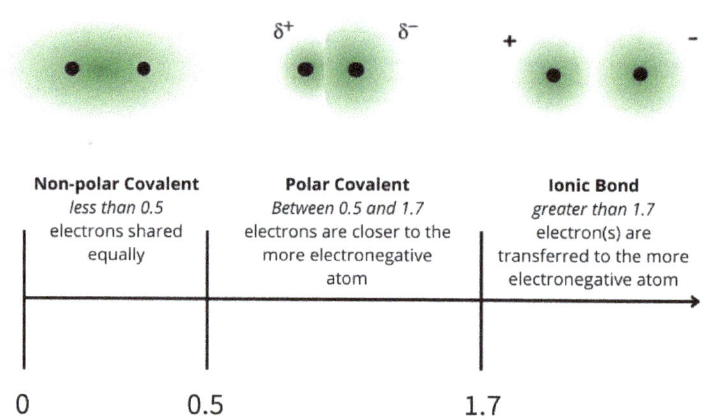

Using Electronegativity to Determine the Bond Type

1. **Identify the electronegativity values:** Look up the electronegativity values of the atoms involved in the bond. Electronegativity values can be found on the periodic table.
2. **Calculate the electronegativity difference:** Subtract the electronegativity value of the atom with lower electronegativity from the atom with higher electronegativity. This will give you the electronegativity difference.

3. **Determine the bond type based on the electronegativity difference:**
   - **Nonpolar covalent bond:** If the electronegativity difference is close to zero or very small (less than 0.5), the bond is considered nonpolar covalent. In nonpolar covalent bonds, electrons are shared equally between the atoms. Examples include C-C or C-H bonds.
   - **Polar covalent bond:** If the electronegativity difference is moderate (between 0.5 and 1.7), the bond is considered polar covalent. In polar covalent bonds, electrons are shared unequally, resulting in a partial positive charge on the less electronegative atom and a partial negative charge on the more electronegative atom. Examples include C-O or C-N bonds.
   - **Ionic bond:** If the electronegativity difference is large (greater than 1.7), the bond is considered ionic. In ionic bonds, electrons are transferred from one atom to another, resulting in the formation of positively charged cations and negatively charged anions. Examples include Na-Cl or K-Br bonds

1. Let's practice! Use the steps above to determine what type of bond would occur between the following atoms:

   C-H

   C-O

   O=O

   Cl-H

Ionic Bonds

In an **ionic bond**, one or more electrons are transferred from one atom to another. This results in the formation of ions, which are atoms or molecules that have gained or lost electrons and, therefore, have a net electrical charge. The atom that loses an electron becomes positively charged, while the atom that gains an electron becomes negatively charged. These opposite charges attract each other and form an ionic bond. This is an important concept that will come up again: atoms or molecules with opposite charges are attracted to each other. Conversely, similarly charged atoms or molecules repel each other.

For example, let's consider the element sodium (Na) and the element chlorine (Cl). Sodium has 1 valence electron, while chlorine has 7 valence electrons. If Sodium were to lose its 1 valence electron, its valence shell would be complete because the next shell in would become the valence shell. Chlorine would need to gain 1 electron in order to have a full outer shell. When these two elements come together, the sodium atom gives its valence electron to the chlorine atom. This loss of an electron results in Sodium forming a +1 charged ion, and when Chlorine gains that electron from Sodium, it forms a -1 charged ion.

These oppositely charged ions are attracted to each other, forming an ionic bond between the two atoms. The resulting compound, sodium chloride (NaCl), is a stable molecule because both atoms have achieved a full outer shell of electrons. Ions with opposite charges are drawn together. When they come into contact, it's as if they recognize and align with each other, forming a strong bond similar to how the charger securely connects to the phone. This attraction results in the formation of a stable compound, as seen in Figure 2.4.

Figure 2.4. The Formation of NaCl Through Ionic Bond

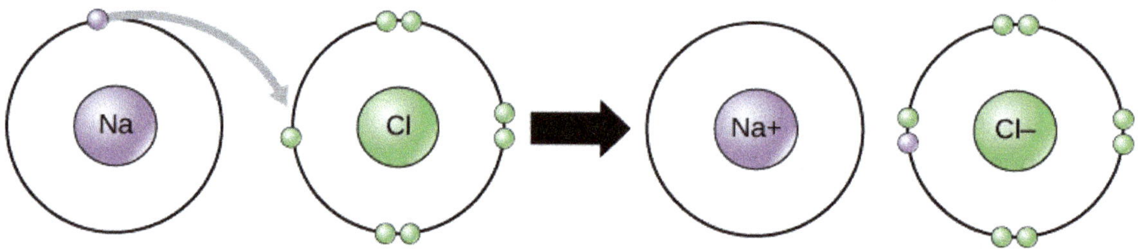

Ionic bonds are relatively strong, and they are important in many biological systems. For example, many enzymes, which are proteins that catalyze biochemical reactions, require specific ions to function properly. However, when ionic substances dissolve in water, like NaCl, the ions are separated. When salt is in water, it exists as Na+ ions and Cl- ions. Additionally, the electrical charges of ions play a role in the maintenance of proper cellular function, as ion channels in cell membranes allow for the controlled movement of ions in and out of cells.

2. Ca^{2+} is an essential nutrient that plays a role in the formation of bones and many other important processes in the body.
   a. How many protons and electrons are found in a neutral Calcium atom?
   b. How many electrons and protons are found in Ca^{2+}?
   c. Keeping in mind that the sum of charges in an ionic compound must always be equal to zero, how many ionic bonds could Calcium make with Chlorine atoms? Explain and draw your response. *Hint: think about how many electrons each atom would donate or receive, and what their charges would be afterward.*

3. Do you think ionic bonds are common in the human body? Explain your answer.

Covalent Bonds

In a covalent bond, two atoms share electrons in order to fill their valence electron shells. This can occur between two atoms of the same element or between two atoms of different elements. Covalent bonds can be polar or nonpolar, depending on how equally the electrons are shared between the two atoms, as seen in Figure 2.5. The type of bond that will occur between two atoms can be determined by calculating the difference in their electronegativity.

Figure 2.5. Types of Covalent Bonds

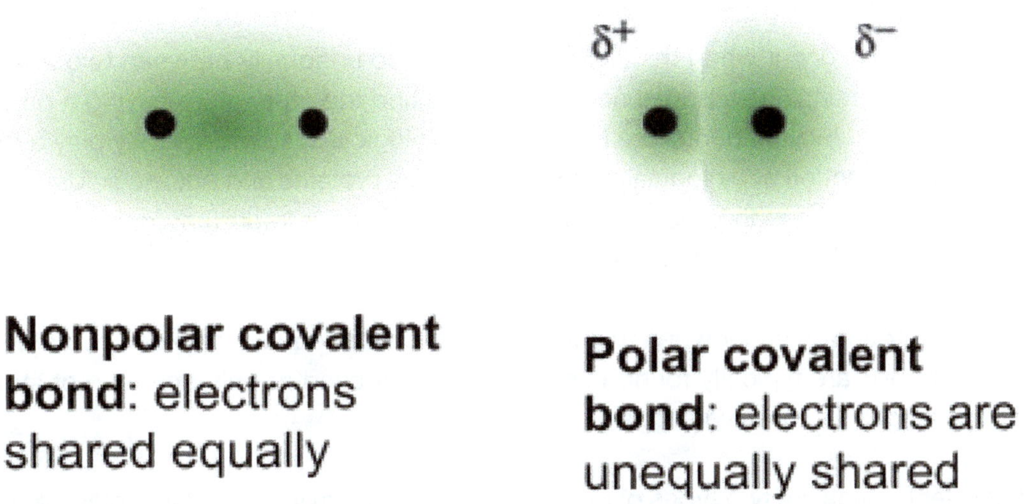

Covalent bonds are represented graphically by a line between two atoms, as shown in Figure 2.6. One line between two atoms means they are sharing two electrons in what's called a single covalent bond. If there is more than one line between two atoms, that means they have made a double or triple covalent bond.

Figure 2.6. Examples of Single and Double Covalent Bonds

(a) A single covalent bond: hydrogen gas (H—H). Two atoms of hydrogen each share their solitary electron in a single covalent bond.

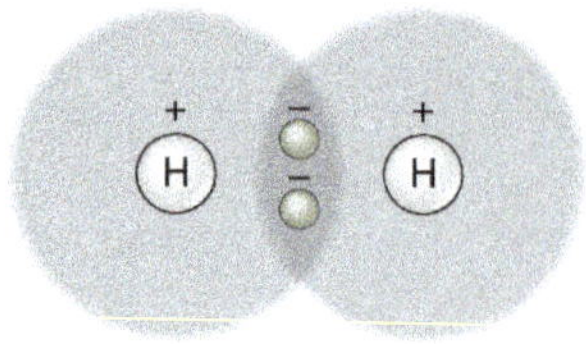

(b) A double covalent bond: oxygen gas (O═O). An atom of oxygen has six electrons in its valence shell; thus, two more would make it stable. Two atoms of oxygen achieve stability by sharing two pairs of electrons in a double covalent bond.

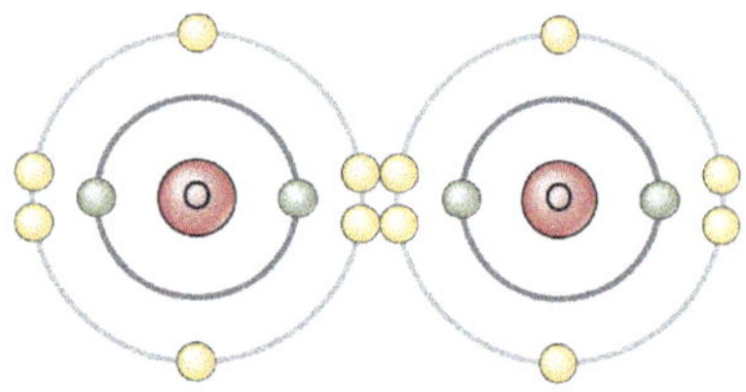

(c) Two double covalent bonds: carbon dioxide (O═C═O). An atom of carbon has four electrons in its valence shell; thus, four more would make it stable. An atom of carbon and two atoms of oxygen achieve stability by sharing two electron pairs each, in two double covalent bonds.

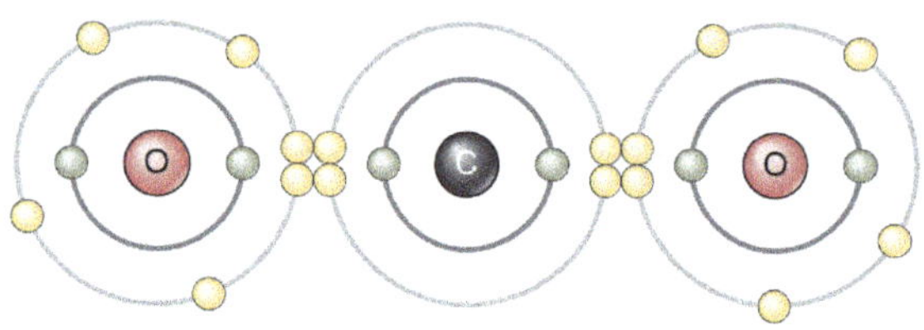

Polar covalent bonds occur when the electrons are shared unequally between two atoms. This results in one atom having a partial negative charge and the other atom having a partial positive charge. Polar covalent bonds are important in biological systems, as they play a role in the structure and function of biological molecules such as proteins and nucleic acids. Polar covalent bonds explain the many characteristics of water, which is important for understanding how cells function since cells are mostly water.

Nonpolar covalent bonds occur when the electrons are shared equally between two atoms. These types of bonds are typically found in molecules composed of only one type of atom, such as hydrogen gas (H_2) or oxygen gas (O_2).

Ionic and covalent bonds are two types of chemical bonds that are important in biological systems. Ionic bonds involve the transfer of electrons between atoms, resulting in the formation of ions, while covalent bonds involve the sharing of electrons between atoms. Understanding the properties and behavior of these types of chemical bonds is important for understanding the structure and function of biological molecules.

4. How are covalent bonds different from ionic bonds?

5. How are the terms electronegativity and polar covalent bond related?

6. How many covalent bonds can Carbon make? Explain how you determined this.

7. How many covalent bonds can Phosphorus make? How do you know?

Hydrogen Bonds

While hydrogen bonds are not true chemical bonds, since electrons aren't interacting, they are a consequence of polar covalent bonds that will be important for understanding the properties of biomolecules like lipids, carbohydrates, and nucleic acids. Polar covalent bonds happen when electrons are unevenly shared between two atoms due to the large difference in electronegativity. This means that electrons will be closer to one atom than another. The atom which has a higher electronegativity will keep the electrons closer. Since electrons are negatively charged, the atom with higher electronegativity will also take on a partial negative charge. The atom with lower electronegativity will take on a partial positive charge since the electrons are slightly farther away from that atom. We denote these partial charges within a polar covalent bond using the lowercase Greek letter delta, which looks like this: δ.

Hydrogen bonds are slight attractions between oppositely charged regions of polar molecules. They are the weakest of all bonds but are important in the structure of many biomolecules.

Let's use methanol and water as an example hydrogen bonds. The molecules are in Figure 2.7.

1. Using the process outlined above, highlight any polar bonds in the molecule.
2. Next, write a δ+ next to the less electronegative atom in the polar bonds, and a δ- next to the more electronegative atom.
3. If the bond is non-polar, you don't need to indicate anything on that region.

Figure 2.7. Methanol and Water

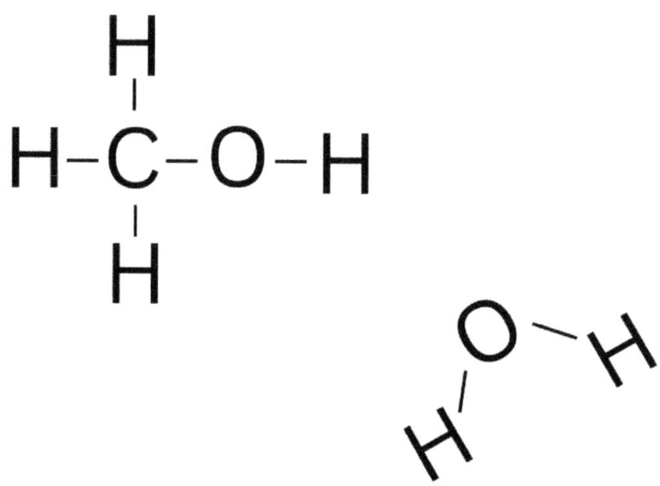

4. Next, let's draw in a hydrogen bond. Draw at least one dashed line between an atom of the water molecule and an atom of the methanol molecule which have opposite charges

8. How are hydrogen bonds different from ionic or covalent bonds?

9. Can nonpolar molecules make hydrogen bonds? Why or why not?

2.3 Reading Chemical Formulas

Now that we know *how* atoms will make bonds to form molecules, let's learn about how to read the formulas of molecules and chemical reactions.

Molecular Formulas

In the example above with hydrogen bonds, we used methanol and water. If we were to write out these molecules in their molecular formula, they would look what you see in Figure 2.8.

Figure 2.8. Structural and Molecular Formula of Methanol and Water

The molecular formula tells us what elements are in that molecule; for example, the H in the CH_4O tells us that there is at least one Hydrogen atom in the methanol molecule. The subscript under the H tells us that there are four hydrogen atoms in one molecule of methanol. But what about Carbon and Oxygen? There isn't a subscript below these letters, but these atoms are present in methanol. If there is no number below the element's symbol, there is only one atom of that element in the molecule. So, under the C and O, it's blank because there is only one Carbon and one Oxygen atom within a molecule of methanol. The "1" is implied but not written.

1. Give the number of atoms of each element in the molecules below:

   C_2H_6

   CO

   $C_3H_8O_3$

Chemical Equations

When molecules react, chemical bonds are broken, and new ones are made, we say a chemical reaction has occurred. We can write out these chemical reactions using chemical formulas which show us what molecules went into the reaction (the reactants) and what molecules came out of the reaction (the products). We'll use the example of methane, CH_4, a gas, and Oxygen gas, O_2, reacting. This is a common reaction called a combustion reaction, which happens in our cells and in gas-powered vehicles.

Figure 2.9. Combustion Reaction of Methane with Oxygen

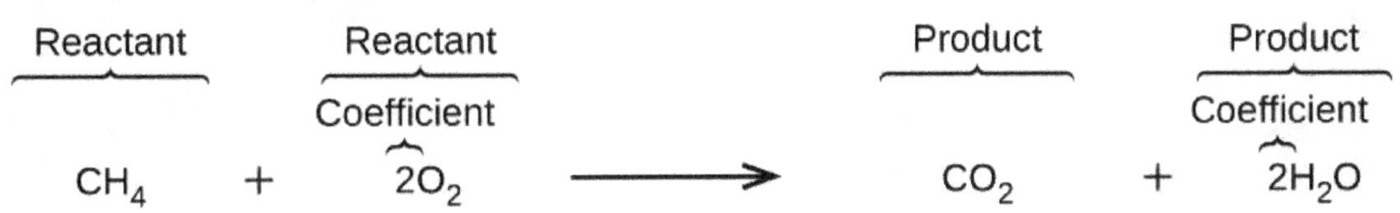

In Figure 2.9, we see another number in the mix, the coefficient. The number that's in front of the molecule tells us how many of that molecule type participate in the reaction. In this example, we have two molecules of Oxygen gas that react with one molecule of methane. Just like with the subscripts, if there isn't a number written in front of the molecule, we can assume the coefficient is simply 1. Both methane and oxygen are reactants—notice how they're on the left side of the arrow? After this chemical reaction is done, the bonds in methane and oxygen have been broken, and we get new substances formed. In this case, we make carbon dioxide, CO_2 and water, H_2O. Everything to the right of the arrow is a product of the chemical reaction. Figure 2.10 is another way of visualizing this reaction.

Figure 2.10. Space Filling Model of the Combustion Reaction of Methane with Oxygen

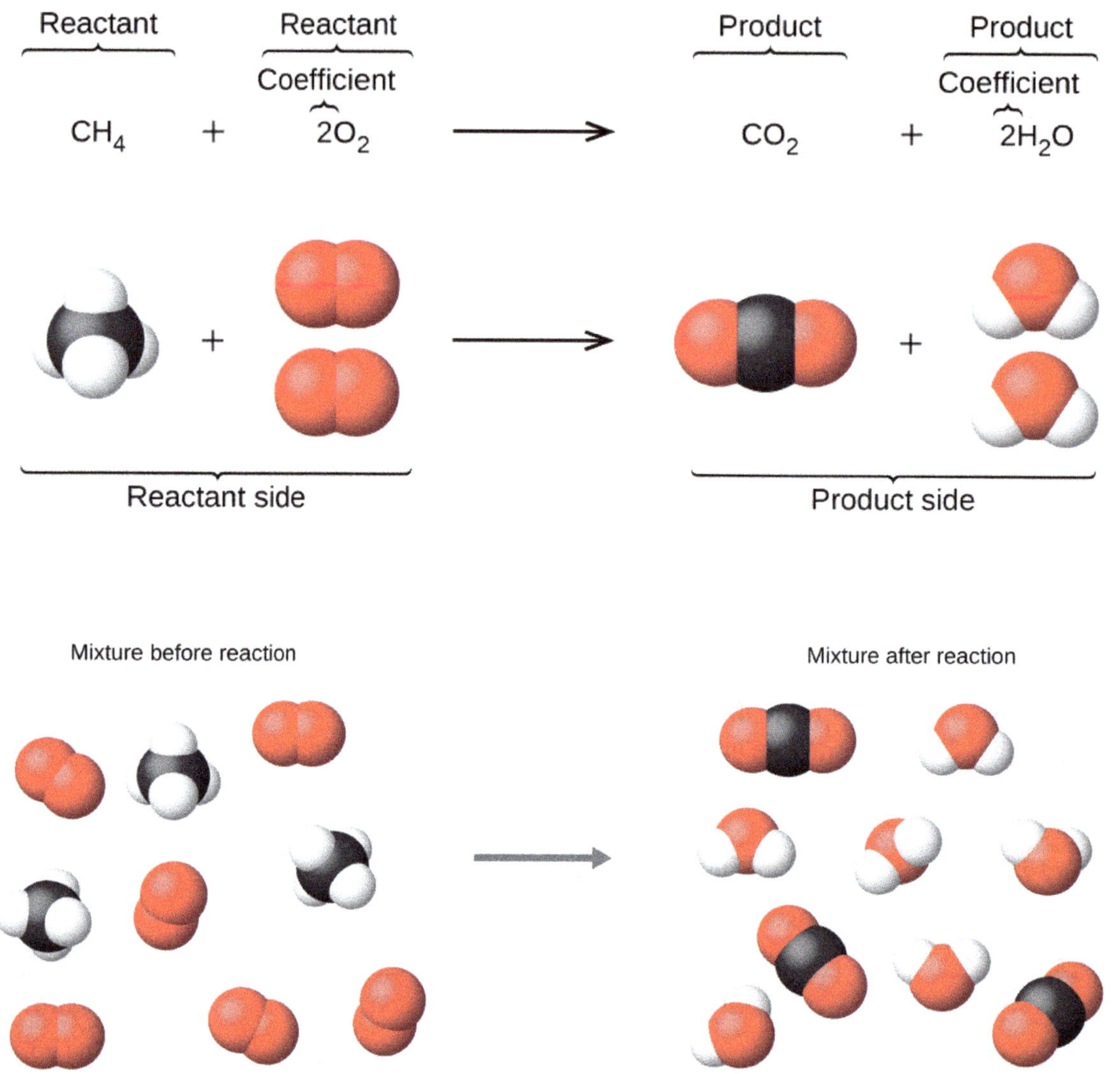

2. Circle the reactants and put a square around the products in the equation below.

$$4\ NO + 2\ O_2 \longrightarrow 4\ NO_2$$

3. Determine the number of each type of atom on the reactants side in the equation above. List them here.

4. Determine the number of each type of atom on the products side in the equation above. List them here.

5. Is the equation above balanced? Explain.

2.4 Chemical Reactions

You've likely seen a chemical reaction occur before. Watching a piece of firewood burn is a chemical reaction. Baking usually involves chemical reactions that transform the raw ingredients of flour, eggs, sugar, and baking powder into something tasty like muffins. We can tell a chemical reaction has occurred when a new substance is formed.

Atoms within a molecule may break and form new bonds with different atoms for many different reasons. We'll talk in a future week about enzymes, which carry out chemical reactions in cells, but for now, let's just get comfortable recognizing when a chemical reaction has occurred and how to read a chemical formula and equation. We'll also introduce a few common types of chemical reactions that we'll see throughout our class.

If we imagine the atoms or molecules in a reaction as words, that may make this introduction to chemical reactions a little easier. The three types of reactions we want to introduce now are synthesis, decomposition (sometimes called hydrolysis), and exchange reactions, as shown in Figure 2.11.

Figure 2.11. Fundamental Reactions in Biology

a) In a synthesis reaction, two components bond to make a larger molecule. Energy is required and is stored in the bond:

NOTE + BOOK → NOTEBOOK

b) In a decomposition reaction, bonds between components of a larger molecule are broken, resulting in smaller products:

BOOKWORM → BOOK + WORM

c) In an exchange reaction, bonds are both formed and broken such that the components of the reactants are rearranged:

NOTEBOOK + WORM → NOTE + BOOKWORM

1. What type of reaction is shown in Figure 2.12? How can you tell?

Figure 2.12. Formation of Sucrose and Barium Carbonate

Glucose Fructose Sucrose

$C_6H_{12}O_6$ $C_6H_{12}O_6$ $C_{12}H_{22}O_5$

$$BaCl_2 + K_2CO_3 \longrightarrow BaCO_3 + 2KCl$$

2. Based on the text, how can we determine that a chemical reaction has occurred, and what are some everyday examples of chemical reactions mentioned?

3. Explain the analogy of atoms or molecules in a reaction being like words. How does this help in understanding chemical reactions, specifically synthesis, decomposition, and exchange reactions?

4. What role do enzymes play in chemical reactions within cells, as briefly mentioned in the text, and why is it important to recognize when a chemical reaction has taken place?

Multiple Choice Review

1. Which type of bond is formed when two atoms share electrons?
   a. Ionic bond
   b. Covalent bond
   c. Hydrogen bond
   d. Metallic bond

2. What characterizes an ionic bond between atoms?
   a. The mutual sharing of electrons between atoms
   b. The transfer of electrons from one atom to another
   c. The weak attraction between polar molecules
   d. The sharing of a free electron cloud in metals

3. How can you predict that a bond between two atoms will be ionic?
   a. If the atoms have similar electronegativities
   b. If one atom has a much higher electronegativity than the other, leading to electron transfer
   c. If the atoms can share electrons equally
   d. If the atoms are both metals

4. What factor is most important in determining whether a bond will be covalent?
   a. The atoms are of the same element.
   b. The atoms have very different atomic sizes.
   c. The atoms share electrons because they have similar electronegativities.
   d. The bond forms between a metal and a nonmetal.

5. Why are valence electrons crucial in determining the chemical properties of an element?
   a. They are involved in the formation of nuclei.
   b. They determine the atomic mass of the element.
   c. They are involved in forming bonds with other atoms.
   d. They are the innermost electrons and thus control all chemical reactions.

6. How do valence electrons affect the reactivity of an element?
   a. Elements with a full valence electron shell are usually more reactive.
   b. Elements with fewer valence electrons are less likely to form bonds.
   c. Elements that have a nearly full or nearly empty valence shell are typically more reactive.
   d. The number of valence electrons has no impact on an element's reactivity.

7. How do atoms in Group 1A typically form bonds with other elements?
   a. By sharing their single valence electron to make covalent bonds
   b. By transferring their single valence electron to another atom, forming an ionic bond
   c. By gaining additional electrons to fill their valence shell
   d. By forming metallic bonds with other Group 1A elements

8. Which of the following elements will have chemical properties most similar to chlorine (Cl)?
   a. Fluorine (F)
   b. Bromine (Br)
   c. Oxygen (O)
   d. Sulfur (S)

3.1 The Properties of Water

A water molecule is made up of two hydrogen atoms bonded to an oxygen atom. Each end of a water molecule has a slight electric charge since it is made up of polar covalent bonds. A molecule that has electrically charged areas is called a **polar molecule**. This uneven distribution of charges across a molecule, making one end positive (H) and the other negative (O), is called **polarity**. The positive hydrogen ends of one water molecule attract the negative oxygen ends of nearby water molecules, causing them to stick together like weak magnets. This attraction causes water molecules to form temporary bonds that break easily. They are called hydrogen bonds, as seen in Figure 3.1.

Figure 3.1. Hydrogen Bonds Between Neighboring Water Molecules

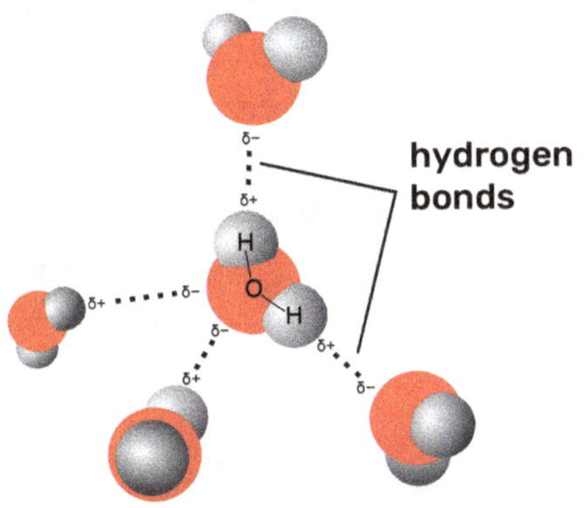

Many of water's important properties occur because of the attraction among its polar molecules. The properties of water include cohesion, adhesion, capillary action, surface tension, the ability to dissolve many substances, and high specific heat.

Figure 3.2. Surface Tension Caused by Water Making Hydrogen Bonds

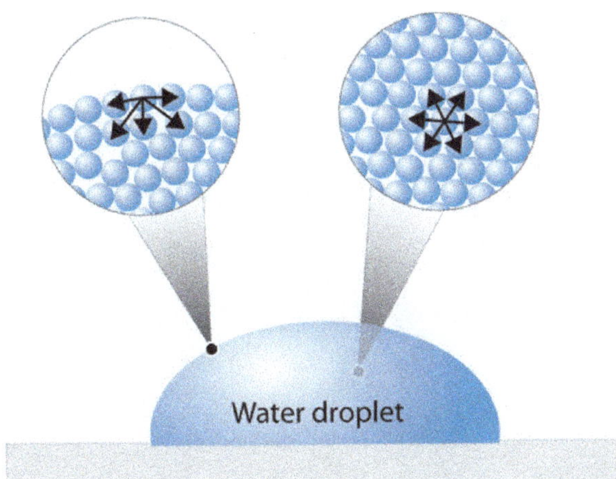

The tendency for water molecules to form weak bonds and stick to each other is called **cohesion**, as seen in Figure 3.2. Because of cohesion, water molecules remain joined together as they move within or between the cells of organisms. A special example of cohesion is surface tension. **Surface tension** is a force that acts on the particles at the surface of a liquid. Surface tension is the tightness across the surface of water that is caused by polar molecules pulling on each other. In liquid water, each water molecule is pulled in all directions by other water molecules. At the surface of the water, however, the attractive force of other water molecules pulls only downward and sideways. This force causes molecules at the surface to be held more tightly together, forming a kind of skin at the water's surface. Small insects, such as water striders, can walk on water by taking advantage of this surface tension. It is what causes raindrops to form round beads when they fall onto a car windshield.

Water molecules are not attracted only to each other. **Adhesion** is the tendency of water to stick to other substances. You see adhesion at work when you add water to a graduated cylinder. At the surface, water creeps up slightly at the sides of the cylinder, forming a curved surface called a **meniscus**. Adhesion allows water to stick to the sides of blood vessels or to the vascular tubes in plants. Both adhesion and cohesion allow water to move in one continuous column from a plant's roots to its leaves. This upward movement, called **capillary action**, as seen in Figure 3.3, is the combined force of attraction among water molecules and with the molecules of surrounding materials, causing a liquid to climb upward against the force of gravity. It allows water to move through materials with pores inside. It also causes water molecules to cling to the fibers of materials like paper and cloth. Capillary action along particular cloth fibers pulls water away from your skin, keeping you dry.

Figure 3.3. Capillary Action of Water

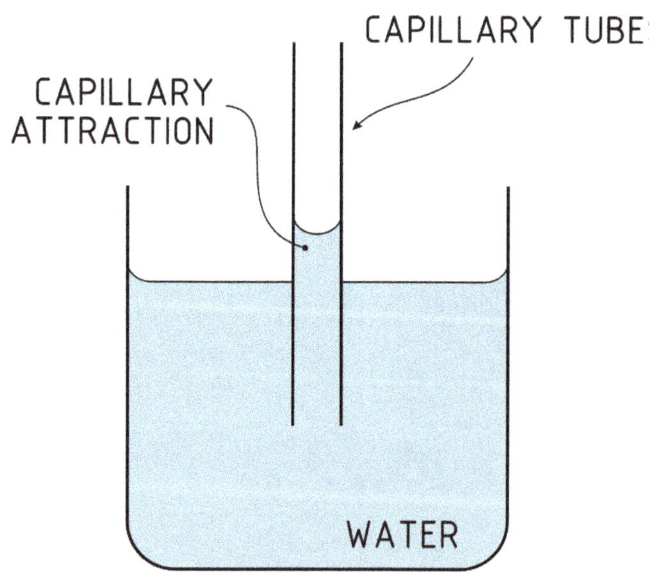

Figure 3.4. Water Dissolving an Ionic Compound, NaCl

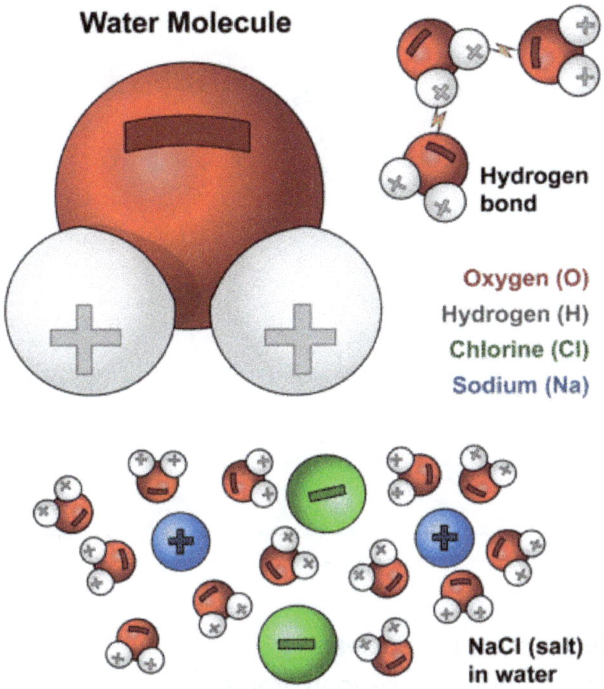

A solution is a mixture that forms when one substance dissolves another. The substance that does the dissolving is called the **solvent**. Water is called the universal solvent because it can dissolve more substances than any other known substance. The main property of water that makes it such a good solvent is its polarity, as seen in Figure 3.4. The charged ends of the water molecule attract the molecules of other polar substances (like dissolves like). It can dissolve substances like sugar, bleach, and salt. It can dissolve gases such as carbon dioxide and oxygen. Dissolving salt is an example. As the negative side of water forms hydrogen bonds with the Na+ and the positive side is attracted to Cl- ions in a crystal of salt (NaCl), it pulls these ions into solution, and the crystal dissolves. The ability of water to dissolve many substances allows water to deliver essential nutrients to cells in plants, animals, and other organisms. Water dissolves nutrients in our food. It does not dissolve nonpolar substances like oil and wax. These molecules do not mix with water and are called **hydrophobic** (fear water).

Specific heat is the amount of heat needed to increase the temperature of 1 g of a substance by 1°C. Compared to other substances, water requires a lot of heat to increase its temperature. The unit of specific heat is joule per kilogram per degree Celsius (J/g •°C). The specific heat of water is very high: 4,184 J/g • °C. Therefore, water takes a long time to heat up or cool down. Water has a high specific heat because of the strong attraction among water molecules (cohesion). This property of water allows lakes, streams, and ocean ecosystems to maintain stable temperatures, even if air temperatures change dramatically. So that the air over water is cooler than the air over land on hot days. The high specific heat of water also helps your body to maintain heat homeostasis.

3

1. Write the definition of each key term.

   Polarity:

   Cohesion:

   Adhesion:

   Capillary Action:

   Surface Tension:

   Specific Heat:

   Hydrogen Bond:

2. Circle all that are true about water.
   a. Water is made up of atoms bonded to form molecules.
   b. Water contains half as many hydrogen atoms as oxygen atoms.
   c. Water molecules tend to push away (repel) each other.
   d. The oxygen in water has a slight negative charge due to its polar bonds with hydrogen.

3. One side of the water molecule has a positive charge, while the other side has a negative charge. What do the charges indicate about the molecule?
   a. Water is a polar molecule.
   b. Water is an ionic compound.
   c. Water is a nonpolar molecule.
   d. Water is an ion.

4. Bonds that form between water molecules are called ____________________ bonds.

5. True or false? Hydrogen bonds are strong and require more energy to break than covalent bonds.

6. The tendency for water molecules to stick together to other water molecules is called ______________________________.

7. How does surface tension force the surface of water to curve?

8. The tendency for water molecules to be attracted and stick to other substances is called ______________________________.

9. Explain how capillary action happens.

10. A mixture that forms when one substance dissolves in another is called a(n) ______________________________.

11. The substance that does the dissolving is called a(n) ______________________.

12. How does water dissolve other substances?

13. A substance that cannot dissolve in water is called ___________________________.

14. The amount of energy needed to increase the temperature of 1g of a substance by 1°C is its ______________________________.

15. True or false? Compared to other substances, water requires a large amount of energy to increase its temperature. ______________________________

3

16. How does the high specific heat of water affect living organisms?

17. Match the terms in Table 3.1 to their definitions.

Table 3.1.

	Cohesion	a. Tendency to stick to other molecules
	Capillary action	b. Weak bonds formed between water molecules
	Specific heat	c. Tendency to stick to other water molecules
	Surface tension	d. Uneven distribution of charges within a molecule
	adhesion	e. Tendency to climb up due to cohesion and adhesion
	Hydrogen bond	f. Tightness caused by the pulling of water molecules on each other
	Polarity	g. The amount of heat needed to raise 1g of a substance by 1

18. Your study partner tells you that all of water's properties that are important to life can be attributed to water's polarity. Do you think this is true? Explain why or why not.

3.2 Acids and Bases

What is pH?

pH is a measurement of how acidic or how basic a solution is. The pH scale starts at 0 and goes up to 14. Halfway between 0 and 14 is 7, which is neutral. Solutions are **acidic** if they have a pH lower than 7. Solutions with a pH higher than 7 are said to be **basic or alkaline**, as seen in Figure 3.5.

Figure 3.5. The pH Scale

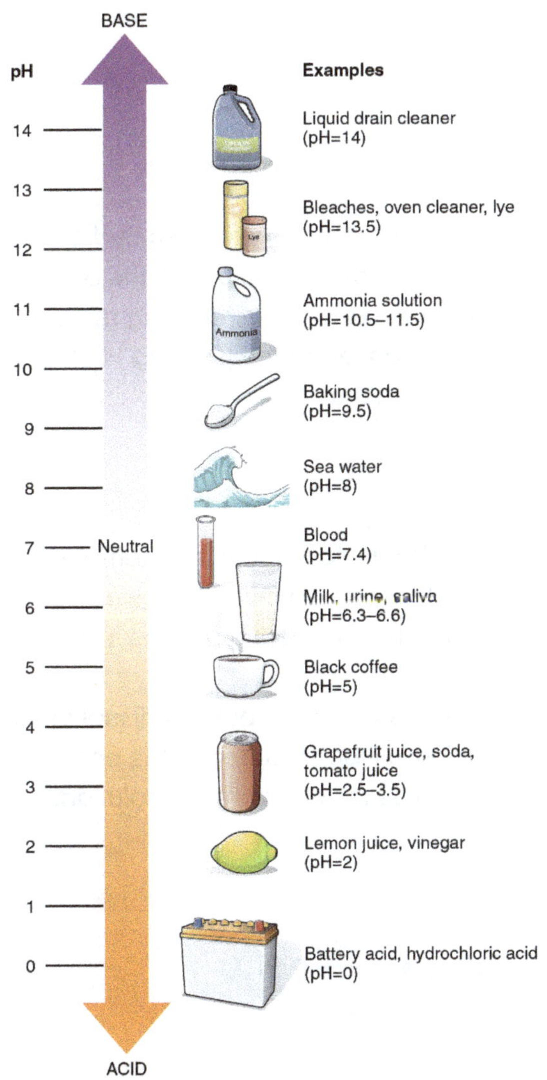

pH means "power of hydrogen" and is a measure of the concentration of hydrogen ions (H^+) in solution. The scale of pH is logarithmic, meaning that a pH of 5 has 10x the concentration of H^+ ions that a pH of 6 has (note the exponents on the image to the right in the H^+ concentration column). The H^+ ion is called a hydrogen ion. It is actually a proton with no electrons.

3

What Makes Something an Acid?

Compounds and molecules are considered an acid if they donate an H^+ ion to solution. This means the H^+ is no longer bound to the molecule it was originally attached to and instead is dissolved in the solution. This increases the concentration of H^+ ions in the solution, making the solution more acidic. Acids add H^+ ions to the solution, therefore increasing the concentration of H^+ ions.

In Figure 3.6, we see hydrochloric acid (HCl). When HCl is dissolved in water, the bond between H and Cl is broken, which separates these two atoms into their ions, H^+ and Cl^-. Because dissolving HCl in water adds H^+ to the solution, HCl is an acid.

Figure 3.6. Dissocation of HCl in Water

$$HCl \longrightarrow H^+ + Cl^-$$

When H^+ ions are in water, they will often pair up with a water molecule to form hydronium. This is because water is polar, and the O in water has a slight negative charge, so the positive H^+ is attracted to the slightly negative O. When this happens, we call the resulting molecule hydronium water with an extra H or H_3O^+. You may see H^+ concentration $[H^+]$ and H_3O^+ concentration $[H_3O^+]$ interchanged. These two values are synonymous.

What Makes Something a Base?

A molecule or compound is considered a base when it adds OH^- to solution or it removes H^+ ions from a solution. This means bases can work in a couple different ways. What we need to focus on is that bases have lower H^+ concentration than acids, and have a pH above 7.

The most common way H^+ ions are removed from solution is when the base is dissolved in water, and it will give off an OH^- (hydroxide) ion. The OH^- ion will be attracted to the positively charged H^+ ion and will bond to form water. This then removes H^+ ions from the solution.

Figure 3.7. The Dissociation of NaOH in Water

$$NaOH \longrightarrow Na^+ + OH^- \longrightarrow OH^- + H^+ \longrightarrow H_2O$$

In Figure 3.7, we see the base Sodium Hydroxide (NaOH). When NaOH is dissolved in water, the bond between Na and OH is broken, which separates these two atoms into their ions, Na^+ and OH^-. Because dissolving NaOH in water adds OH^- to the solution and the OH^- then removes H^+ by bonding to it, NaOH is a base. When H^+ ions are bonded to a molecule, they are not contributing to the pH.

Another way bases remove H^+ ions from the solution is to bond with them directly. For example, ammonia, NH_3 will bond to an H^+ ion to form ammonium, NH_4^+, which removes that ion from the solution since it is now bonded to the N in the ammonium molecule. Ammonia can even take an H^+ from a water molecule, which results in an increase of OH^- ions, making the pH basic (higher pH), as seen in Figure 3.8.

Figure 3.8. Ammonia in Water

$$NH_3 + H^+ \longrightarrow NH_4^+$$

$$NH_3 + H_2O \longrightarrow NH_4^+ + OH^-$$

The key here is the bases have a lower H^+ concentration, and therefore have a high OH^- concentration. How this happens depends on the type of molecule. What we need to know is that bases have a pH above 7, have a higher OH^- concentration than H^+ concentration, and usually donate OH^- to solution.

1. What does pH measure?

2. What is an ion?

3. What is a hydroxide ion?

3

4. What is a hydrogen ion?

5. What is the hydronium ion and its formula?

6. What makes something an acid?

7. What makes something a base?

8. Acids have ____________ (high/low) H^+ concentration and a pH range of _________________.

9. Bases have ____________ (high/low) H^+ concentration and a pH range of _________________.

How Do We Measure pH?

The pH scale is a measure of the concentration of Hydrogen ions. The units used in the pH scale are Molar (or Molarity). Molar units are the number of moles of a substance per liter of solution. So, if we have 1 mole of Hydrogen ions dissolved in 1L of solution, the Molar concentration would be 1 Molar, or 1M, as seen in Figure 3.9.

3

Figure 3.9. Molar Concentration Equation

$$\text{Molar} = \frac{\text{moles}}{\text{Liter}}$$

We represent Hydrogen ion concentration by putting the name of the ion in brackets. For example, if we have 0.01M H^+ concentration, we would write it as 0.01M = $[H^+]$.

Figure 3.10. The Relationship Between pH and H+ ion Concentration

$$pH = -\log[H^+]$$

Since pH is a logarithmic scale, each step on the pH scale represents a 10x change in concentration of H^+ ions. To calculate pH from $[H^+]$, we use the equation in Figure 3.10. If we have $[H^+] = 1.0 \times 10^{-5}$, the pH would be 5.

Figure 3.11. The Relationship Between H+ ion Concentration and pH

$$10^{-ph} = [H^+]$$

If we rearrange the equation, as seen in Figure 3.11, we can figure out $[H^+]$ from the pH. For example, if the pH is 8, the $[H^+]$ would be 1.0×10^{-8}.

10. What is the pH of a solution with $[H^+] = 1.0 \times 10^{-2}$?

11. Which has a higher [H^+], pH of 3 or pH of 2? *Hint: write out the [H^+] in decimal, not scientific notation to help visualize this.*

3

Water Can Act as a Base or an Acid

To understand what this means, you must understand water. Water is a molecule made up of three atoms covalently bonded together. Think of water as HOH.

Sometimes water molecules break apart into H^+ and OH^- ions. This happens even in pure water, but in pure water the concentration of H^+ and OH^- are equal, so the solution has a pH of 7 (neutral). Remember hydronium above? The extra H^+ ion that forms when water dissociates can be picked up by another water molecule to form hydronium, as shown below. This means water can both donate H^+ ions, and take up H^+ ions, so it can act as both an acid and a base, as seen in Figure 3.12.

Figure 3.12. Autoionization of Water

$$H_2O + H_2O \longrightarrow OH^- + H_3O^+$$

Acid-Base Neutralization

If you mixed hydroxide and hydrogen ions together, they would immediately pair up and make water molecules. Each H^+ ion can bond with one OH^- ion to form water, as seen in Figure 3.13.

Figure 3.13. Acid Base Neutralization

$$H^+ + OH^- \longrightarrow H_2O$$

This is called a neutralization reaction. Hydroxide ions neutralize hydrogen ions. If, after the neutralization reaction is complete, there are H^+ ions left over, then the solution is acidic. Or, if after the neutralization reaction is complete, there are OH^- ions left over, then the solution is basic.

12. What is the chemical formula for water?

13. Name the two ions that form when water dissociates.

14. What ion is needed to neutralize an acid?

15. What ion is needed to neutralize a base?

16. What forms when an acid and a base neutralize each other?

pH Homeostasis & Buffers

Most cells can only survive within a certain range of pH. Different parts of the body maintain slightly different pH levels, based on their function. The pH level of any part of the body, and the cytoplasm (inside region) of the cell is maintained by molecules called buffers.

Buffers are molecules that minimize changes in pH that are caused by normal everyday activities of organisms and cells. *Buffers can neutralize the addition of both acids and bases to a solution.* Each type of buffer has a different molecular formula and works to keep the pH at a certain level—not all buffers maintain neutral pH. For instance, in your stomach, buffers work to keep that pH around 2, as seen in Figure 3.14.

But why does our body need to maintain a constant pH? *Because homeostasis is life*. On a more granular level, acids denature, or change the shape of proteins in much the same way heat does. As a matter of fact, strong acids like vinegar and lemon juice can be used to actually "cook" meats like fish and eggs. Ceviche is a dish made by mixing raw fish and lime juice and letting it sit for a few hours. Acids are used by your digestive system to break down food molecules into simpler monomers that can then be absorbed into the bloodstream. Bases (that have a high OH^- concentration) typically impact the structure of lipids, which make up the cell membrane. This is why if you spill bleach on your hand, it doesn't burn but feels soapy, that's the lipids in your cell membranes being broken down.

Figure 3.14. pH of Various Parts of the Human Body

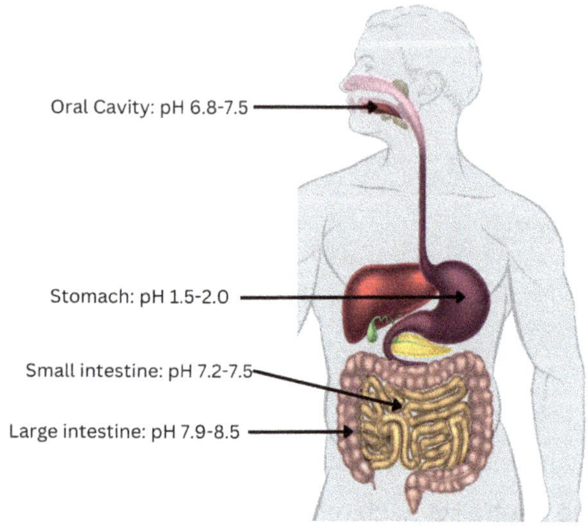

Our bodies maintain near-constant pH by producing buffers. These buffers maintain pH by accepting or donating H^+ ions when the pH of the solution changes. This helps proteins work properly without being denatured (unfolded) and is one way our bodies maintain homeostasis, as seen in Figure 3.15.

Figure 3.15. Buffers Mode of Action

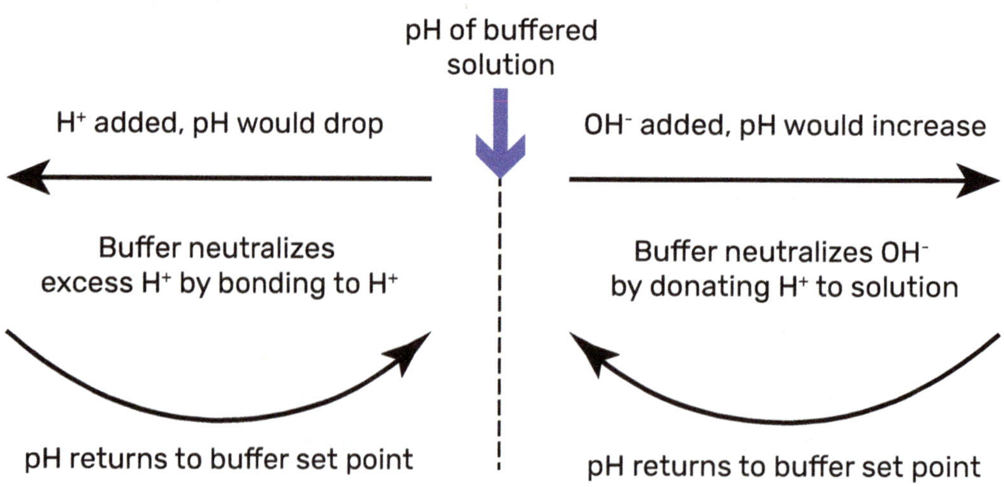

Buffers in our blood

The reaction you need to know to understand blood function in Anatomy and Physiology is how carbonic acid and bicarbonate works in our blood. For example, human blood has a pH of about 7.2, which is slightly basic. Any higher or lower and the blood cells would be damaged. A healthy person's blood has a pH range or 7.2 to 6.9. We'll use this as an example of how buffers work. You should know that buffers maintain pH by adding H^+ ions when a base (OH^-) is added, and bonding to H^+ ions when acids are added.

During the normal process of breathing, carbon dioxide is dissolved into our blood. Our blood is mostly water, and when carbon dioxide (CO_2) is dissolved in water, it forms carbonic acid. Carbonic acid is a weak acid, which means it doesn't easily give up a H^+ ion. But, when carbonic acid does give up an H^+ ion, it forms bicarbonate. Bicarbonate is a buffer because it can either bond to excess H^+ ions in solutions and go back to its carbonic acid form. In very basic conditions, it can give up another H^+ to form CO_3^{2-}. It's this reaction with bicarbonate that helps maintain a stable blood pH, as seen in Figure 3.16.

Figure 3.16. Carbonic Acid Formation

$$\underbrace{CO_2}_{\text{carbon dioxide}} + \overbrace{H_2O}^{\text{water}} \longrightarrow \underbrace{H_2CO_3}_{\text{carbonic acid}} \longrightarrow \overbrace{H^+}^{\text{hydrogen ion (acid)}} + \underbrace{HCO_3^-}_{\text{bicarbonate (conjugate base)}}$$

17. What happens if you add an acid to a buffered solution?

18. What happens if you add a base to a buffered solution?

19. Why are buffers important to life?

20. What characteristic of life would buffers relate to?

3

Multiple Choice Review

1. What is the pH range of an acidic solution?
   a. 0–7
   b. 7–14
   c. 7
   d. 0–14

2. Which substance is an example of a base?
   a. Vinegar
   b. Lemon juice
   c. Ammonia
   d. Stomach acid

3. What is the pH of a neutral solution, like pure water?
   a. 0
   b. 7
   c. 14
   d. 10

4. How do acids differ from bases in terms of hydrogen ion (H^+) concentration?
   a. Acids have a higher H^+ concentration than bases.
   b. Bases have a higher H^+ concentration than acids.
   c. Acids and bases have the same H^+ concentration.
   d. Acids do not contain H^+ ions.

5. What happens to the pH of a solution if the concentration of hydrogen ions increases?
   a. pH increases
   b. pH decreases
   c. pH remains the same
   d. pH becomes neutral

6. Which of the following best describes a buffer solution?
   a. It changes pH drastically when an acid or base is added.
   b. It prevents pH changes when small amounts of acid or base are added.
   c. It only neutralizes acids, not bases.
   d. It increases the hydrogen ion concentration in solutions.

7. What is the role of buffers in biological systems?
   a. To stop all chemical reactions
   b. To maintain constant pH
   c. To eliminate acids from the body
   d. To increase pH levels

8. How does a base react in water?
   a. It releases hydrogen ions (H^+).
   b. It accepts hydrogen ions (H^+).
   c. It releases hydroxide ions (OH^-).
   d. It accepts hydroxide ions (OH^-).

9. In the pH scale, which value represents a very strong acid?
   a. 1
   b. 7
   c. 10
   d. 13

10. What is the effect of adding a strong acid to a buffered solution?
    a. The pH of the solution significantly increases.
    b. The pH of the solution significantly decreases.
    c. The buffered solution resists a large change in pH.
    d. The buffered solution becomes a strong base.

1.1 Dehydration Synthesis and Hydrolysis Practice

Dehydration synthesis and hydrolysis are fundamental biochemical processes that facilitate the formation and breakdown of biological molecules within cells.

Dehydration synthesis, also known as condensation reaction, is a process where two molecules combine to form a larger molecule, with the removal of water. Dehydration synthesis is how monomers connect to make polymers. In this process, -OH from one molecule and a hydrogen atom (H) from another molecule are removed, resulting in the formation of water (H_2O) and a new covalent bond between the two original molecules. This reaction is key in forming larger biomolecules like carbohydrates, proteins, and nucleic acids.

Hydrolysis is the reverse of dehydration synthesis. It involves breaking down a complex molecule into smaller units by adding water. Hydrolysis is how polymers are broken down into monomers. During this process, a water molecule is split into a -OH and a hydrogen (H) atom, which are then incorporated into the monomers. This reaction is crucial in the digestive system, where complex food substances are broken down into smaller molecules that can be absorbed by cells in the intestine. For instance, in the digestion of carbohydrates, polysaccharides are hydrolyzed into monosaccharides, making them usable for cellular energy.

Both processes are catalyzed by enzymes and are essential for the metabolism and regulation of biomolecules in the cell, playing critical roles in the pathways that synthesize and degrade cellular components.

1. Match the terms in Table 4.1 to their definitions.

Table 4.1.

	Dehydrate	a. To split or break apart
	Hydro-	b. To make something
	Synthesis	c. Many monomers hooked together make a:
	-lysis	d. To lose or remove water; to take water away
	Monomer	e. Means water (as in pairing water
	Polymer	f. Building block or single unit of a polymer

2. Examine each figure. Indicate if each is an example of dehydration synthesis or hydrolysis.

Figure 4.1 Reaction ____________________________________

Figure 4.1.

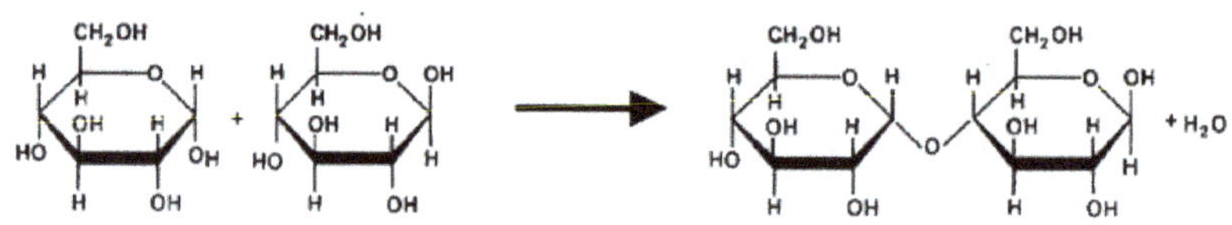

Figure 4.2 Reaction ____________________________________

Figure 4.2.

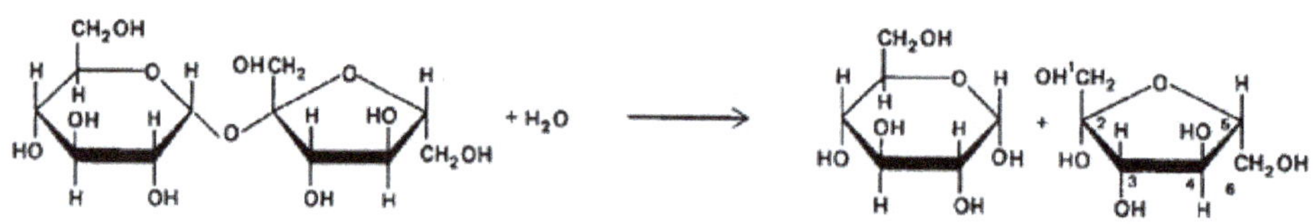

Figure 4.3 Reaction ____________________________________

Figure 4.3.

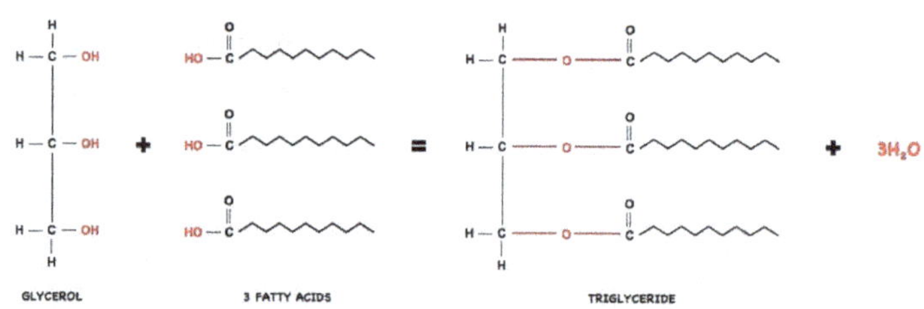

Figure 4.4 Reaction ____________________________________

Figure 4.4.

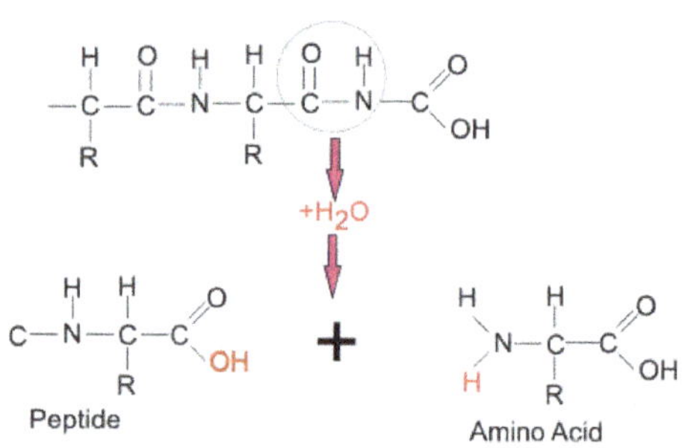

3. Explain in your own words: How can you tell if a chemical equation represents:
   a. Dehydration synthesis?

   b. Hydrolysis?

4

Figure 4.5 is an example of dehydration synthesis. In dehydration synthesis, a hydrogen atom from one molecule joins with a hydroxyl group (-OH) from another molecule to form water, leaving two molecules bonded to the same oxygen atom. For example, when glucose and fructose combine by dehydration synthesis, they form sucrose and water. Analyze Figure 4.5 to answer the questions that follow.

Figure 4.5. Dehydration Synthesis

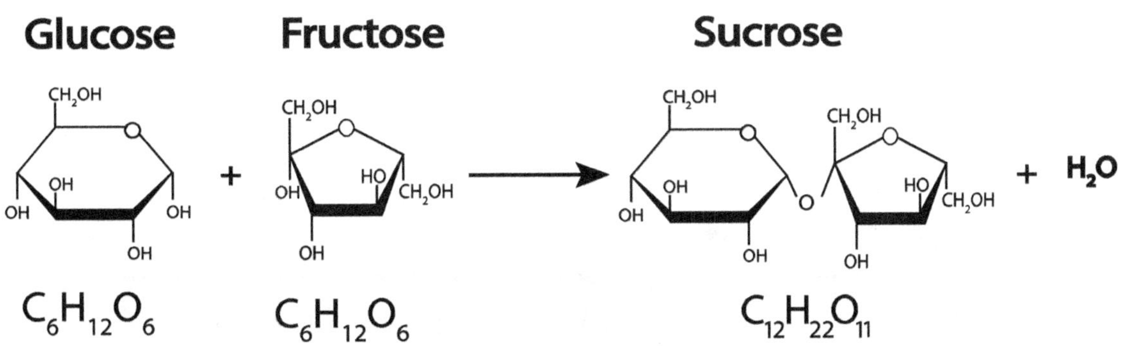

Figure 4.6 is an example of hydrolysis. Complex organic molecules are broken down by the addition of the components of water, - H^+ and OH^-.

Figure 4.6. Hydrolosis

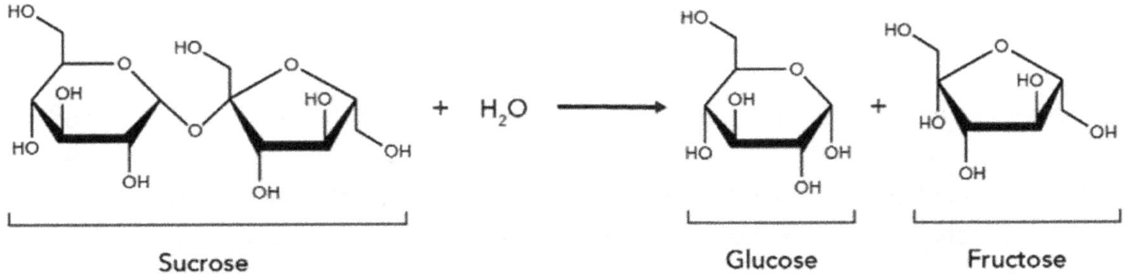

4

4. What are the reactants of the dehydration synthesis reaction?

5. What are the products of the hydrolysis reaction?

6. How are these two reactions related?

Summary Review

1. The JOINING of two monomers causes a water molecule to be lost. This joining to make a polymer is called ____________________

2. The SPLITTING apart of two organic molecules in a polymer and adding back the water parts to make individual monomers again is called ________________________

3. The organic molecules that serve as a source of energy for us are commonly called ______________. In what organ of your body would the splitting apart (hydrolysis) of these polymers into their monomers be occurring at a high rate right now? __________ ______________________

4. How many water molecules are produced when you join 114 amino acids together?

5. During dehydration synthesis if 42 water molecules were made how many monosaccharides were joined together to make the complex carbohydrate?

Multiple Choice Review

1. Hydrolysis can be best described as a process that:
   a. requires the input of energy without water.
   b. releases energy without using water.
   c. consumes water to break bonds.
   d. produces water to make bonds.

2. When forming a disaccharide from two glucose molecules, what is a byproduct?
   a. Carbon dioxide
   b. Water
   c. Oxygen
   d. Hydrochloric acid

3. If a biologist wants to break down a protein into its amino acid components, which process would they use?
   a. Dehydration synthesis
   b. Hydrolysis
   c. Photosynthesis
   d. Fermentation

4. In which scenario would a cell typically use dehydration synthesis?
   a. When breaking down starch into glucose
   b. When converting glucose to glycogen for storage
   c. When digesting protein from food
   d. When absorbing minerals from the environment

5. What role does hydrolysis play in digestion?
   a. It breaks down nutrients into absorbable units by adding water.
   b. It synthesizes new molecules to be absorbed.
   c. It dehydrates the food to preserve it.
   d. It catalyzes the reaction without water.

6. During dehydration synthesis, what happens to the molecules involved?
   a. They lose water and separate.
   b. They gain water and combine.
   c. They lose water and combine.
   d. They gain water and separate.

1.2 Carbohydrates

There are three different groups of carbohydrates: monosaccharides, disaccharides and polysaccharides. "Saccharide" means sugar, from the Latin "saccharum." Most carbohydrates serve the function as a source of readily available energy for cells. Some polysaccharides have additional functions including structural support and cell to cell recognition. Most carbohydrates tend to end in -ose: sucrose, cellulose, lactose, etc.

Monosaccharides

A single molecule sugar is called a monosaccharide. The prefix "mono-" means one. Each monosaccharide has a slightly different structure, though many will have the same chemical formula. Examples of three monosaccharides are shown below: glucose, fructose, and galactose. Monosaccharides will often take a ring shape when they are dissolved in water. Three common monosaccharides are shown in Figure 4.7.

Figure 4.7. Common Monosaccharides

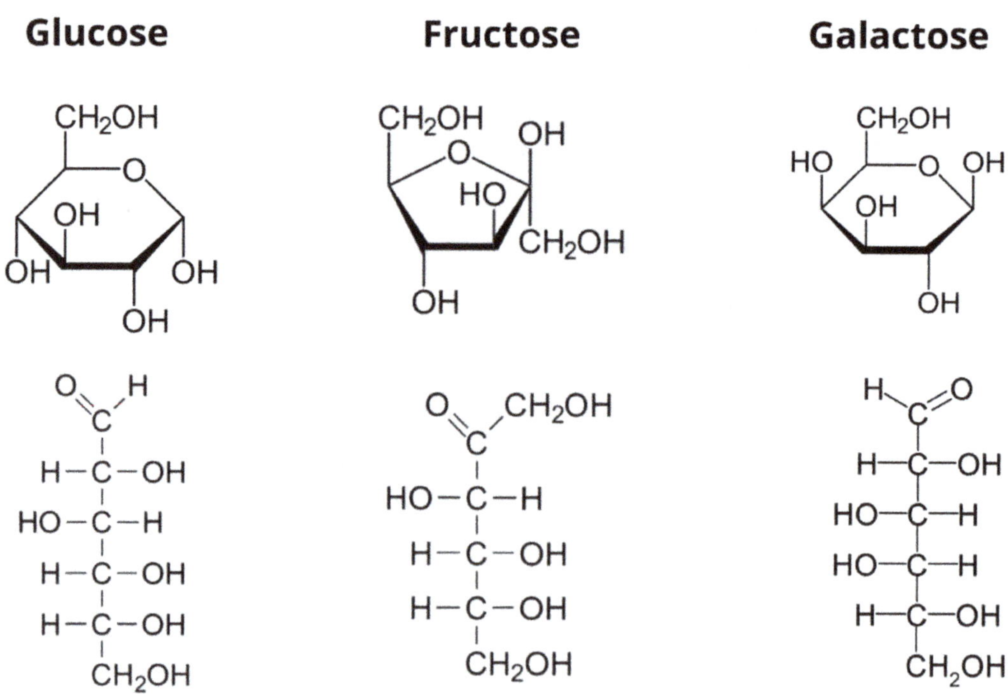

1. What three elements are present in the three monosaccharides shown above?

2. How many atoms of carbon are present in a molecule of:
   a. glucose _______
   b. fructose ________
   c. galactose ________

3. Add subscripts to the following to indicate the proper formula for each molecule. Fill in the blanks by counting the total number of carbon, hydrogen, and oxygen atoms in each molecule.
   a. Glucose C______ H______ O_______
   b. Fructose C______ H______ O_______
   c. Galactose C______ H______ O_______

4. Explain why carbohydrates are often abbreviated as CH_2O and what the name "carbo hydrates" indicates.

5. Compare the structural formula of glucose to fructose. Are they exactly the same shape?

6. Glucose, fructose, and galactose have the same chemical formula but different structures. This makes them ________________

Disaccharides

Two monosaccharide sugar molecules can join together to form a larger carbohydrate molecule called a disaccharide. The prefix "di-" means two. Dehydration synthesis is a reaction in which two molecules become covalently bonded to each other through the release of a water molecule. Hydrolysis is a chemical process that reverses dehydration synthesis by the addition of water. Three common disaccharides are shown in Figure 4.8

Figure 4.8. Common Disaccharides

Sucrose
glucose-fructose

Lactose
galactose-glucose

Maltose
glucose-glucose

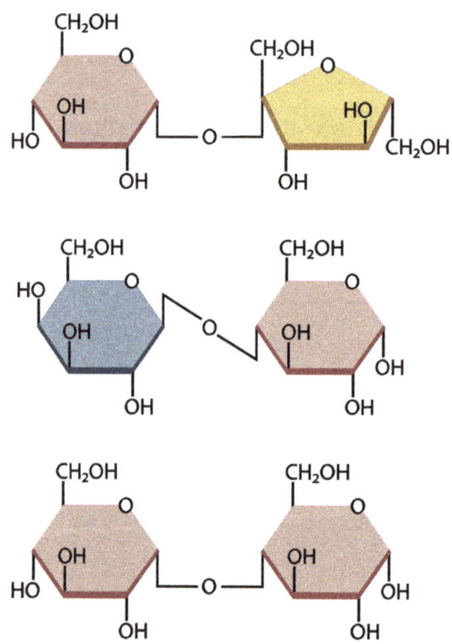

Different disaccharide molecules can be made by joining different monosaccharides in different combinations. By joining together a glucose molecule with another glucose molecule, a disaccharide called maltose is formed. By joining a glucose and a fructose molecule together, sucrose is formed (table sugar). The technical name for the bond between monosaccharide molecules is a glycosidic bond.

7. What is removed from monosaccharides to build a disaccharide?

8. Write a simple formula for maltose.

9. How does the simple formula for sucrose compare to maltose?

10. Why is the formula for sucrose the same as maltose?

11. Why isn't the formula for maltose equal to two glucose molecules?

12. How many monosaccharide molecules are needed to form one sucrose molecule? One maltose molecule?

Polysaccharides

Just as double sugars are formed from two single sugar molecules, polysaccharides are formed when many simple sugars are joined through dehydration synthesis. The prefix "poly-" means many. Starch, glycogen, and cellulose are the three most common polysaccharides in biology. They each consist of long chains of glucose molecules joined. Three common polysaccharides are shown in Figure 4.9.

Figure 4.9. Common Polysaccharides

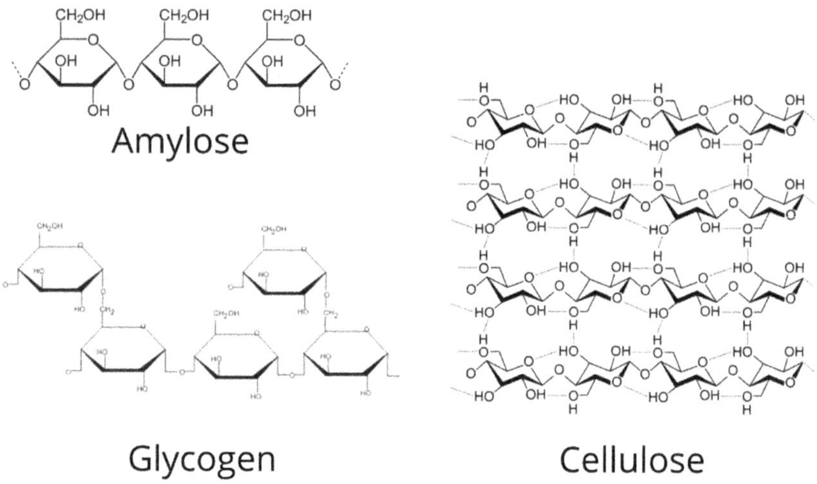

Starch (amylose) is commonly found in plants and is a storage form of glucose. **Glycogen** is commonly found in animal tissues as a storage form of glucose. It's often easier to access glycogen than fat as a source of energy. **Cellulose** is found in plants and is the main structural component of cell walls. Most animals do not possess the enzymes necessary to digest starch to use its glucose as an energy source.

13. Glucose is the primary source of energy for cells. Foods rarely contain monosaccharides, but disaccharides and polysaccharides are common. What process is used in the digestion of disaccharides and polysaccharides to make glucose available to the cells?

14. Cellulose is an extremely long polysaccharide that is very difficult for most animals to digest. Why?

Multiple Choice Review

1. What is the main functional difference between monosaccharides and polysaccharides in biological systems?
   a. Monosaccharides store genetic information, while polysaccharides provide structural support.
   b. Monosaccharides are used for quick energy, while polysaccharides are used for energy storage and structural support.
   c. Monosaccharides are primarily used in cell signaling, while polysaccharides are used in cell division.
   d. Monosaccharides build nucleic acids, while polysaccharides create amino acids.

2. Which of the following is a disaccharide?
   a. Glucose
   b. Starch
   c. Lactose
   d. Cellulose

3. How are monosaccharides related to polysaccharides?
   a. Monosaccharides are individual units that bond together to form polysaccharides.
   b. Polysaccharides are broken down into monosaccharides during digestion.
   c. Both A and B
   d. Neither A nor B

4. What type of bond links individual sugar molecules in disaccharides and polysaccharides?
   a. Hydrogen bond
   b. Synthetic bond
   c. Covalent bond
   d. Ionic bond

5. Which monosaccharide is commonly known as "fruit sugar"?
   a. Glucose
   b. Fructose
   c. Sucrose
   d. Lactose

6. What is the primary function of polysaccharides in plants?
   a. To act as a primary source of energy
   b. To serve as structural components
   c. To facilitate cell communication
   d. To speed up biochemical reactions

7. Which of the following is not a polysaccharide?
   a. Glycogen
   b. Starch
   c. Maltose
   d. Cellulose

8. How do disaccharides form from monosaccharides?
   a. Through a hydrolysis reaction
   b. Through a dehydration synthesis reaction
   c. Through an oxidation-reduction reaction
   d. Through a polymerization process

9. Why are polysaccharides like starch and glycogen important for energy storage?
   a. They are easily transported through the bloodstream.
   b. They provide quick and short-term energy bursts.
   c. They can be broken down into glucose when energy is needed.
   d. They are permanent stores of energy and are never broken down.

10. Which of the following best describes the solubility of carbohydrates in water?
    a. Monosaccharides are generally insoluble in water, while polysaccharides are soluble.
    b. Polysaccharides are generally insoluble in water, while monosaccharides are soluble.
    c. Both monosaccharides and polysaccharides are insoluble in water.
    d. Both monosaccharides and polysaccharides are soluble in water.

1.3 Lipids

Lipids are fats including oil, waxes, steroids, and cholesterol. Each type of lipid has a different molecular structure (and therefore function), but they all share the property of being hydrophobic due to a large percentage of bonds being non-polar C-C and C-H bonds.

Triglycerides

Triglycerides, commonly called fats, are made from a glycerol backbone and fatty acid tails. Fatty acids are a hydrocarbon chain with a carboxyl group connecting it to the glycerol, as seen in Figure 4.10. Glycerol is a three-carbon alcohol. There are two types of fatty acids, saturated and unsaturated. Saturated fatty acids are straight and are found mostly in animals. Unsaturated fatty acids are bent because of a double bond and are found mostly in plants.

Figure 4.10. Synthesis of a Triglyceride

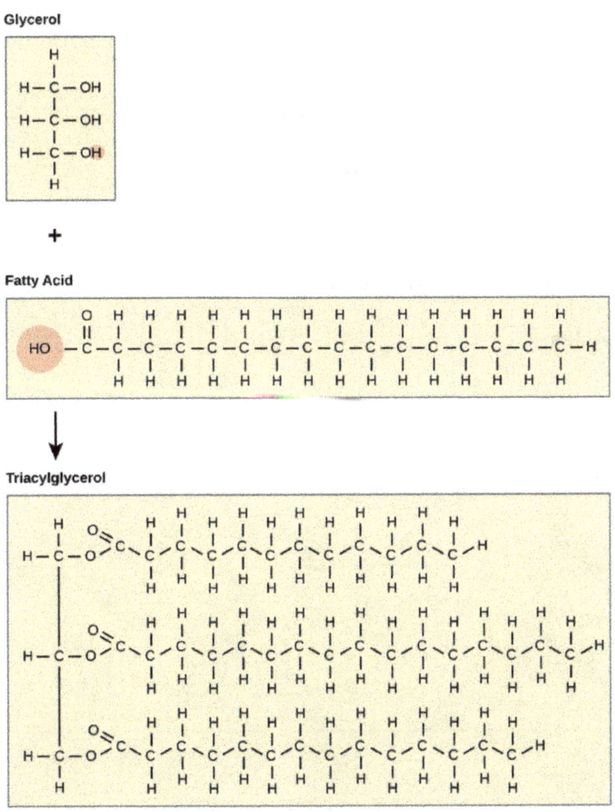

The main function of triglycerides is to store energy in a compact form. Each gram of fat contains ~9 calories of energy, compared to about 4 calories of energy in a gram of carbohydrate.

1. Looking at the fat molecule in Figure 4.11, add in the missing fatty acids.

Figure 4.11. Draw in the Missing Parts of the Tryglyceride

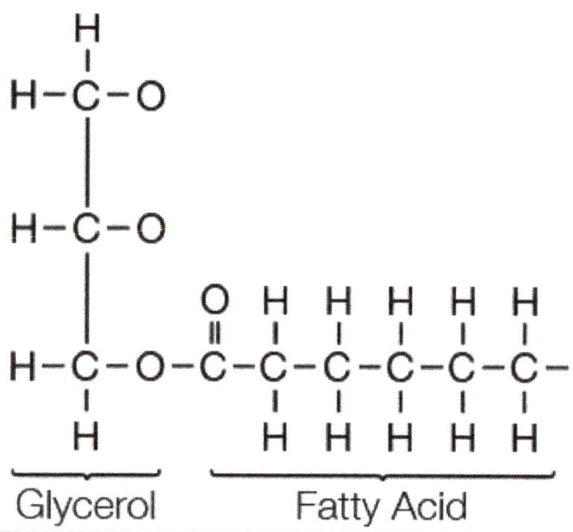

2. Fatty acids are composed of a carboxyl group and ______________________.

3. What are the four parts of a triglyceride? One ________________ and three ________________.

4. The human body stores energy as __________________________ fat.

5. Label the fatty acids in Figure 4.12 as saturated or unsaturated.

Figure 4.12. Saturated and Unsaturated Fatty Acids

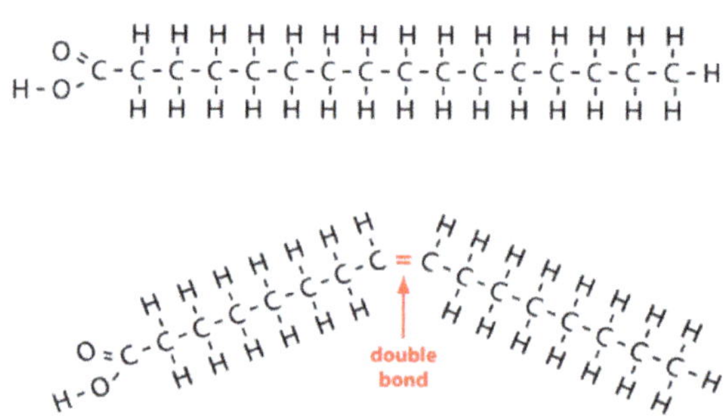

6. How do you think the bend caused by the C=C in unsaturated fats is related to the fact that these molecules tend to be liquid at room temperature?

Phospholipids

Phospholipids are similar to triglycerides except that a phosphate group replaces one of the fatty acids. The phosphate group is a very polar molecule. This gives the phospholipid the unique property of one portion of the molecule being hydrophilic (the phosphate group), and the rest of the molecule being hydrophobic (the fatty acids).

7. Fill in one phosphate group and the remaining fatty acid in Figure 4.13. Label all the parts. One fatty acid should be unsaturated.

Figure 4.13. Draw in the Missing Parts of the Phospholipid

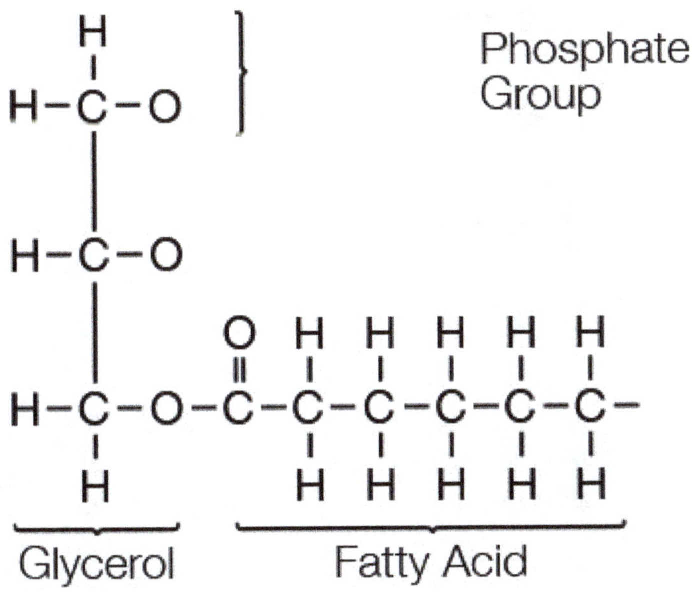

Phospholipids are commonly drawn as a circle with two tails. Phospholipids are polar on the phosphate group side and nonpolar on the fatty acid side. When phospholipids are mixed with water, they line up with heads on one side facing out towards water, and their tails on the inside, away from water. One layer of phospholipids is called a micelle. Two rows of phospholipids are called a lipid bilayer, and they make up the main component of all cell membranes, as diagramed in Figure 4.14.

Figure 4.14. Phospholipids

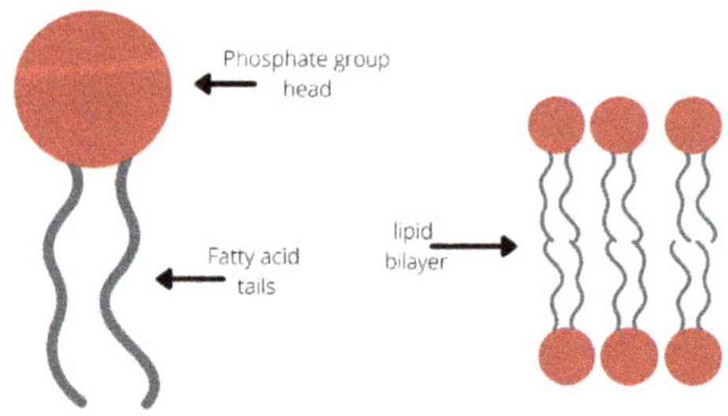

8. Label the regions in the bilayer in Figure 4.15 as polar or nonpolar.

Figure 4.15. Phospholipids Form Bilayers

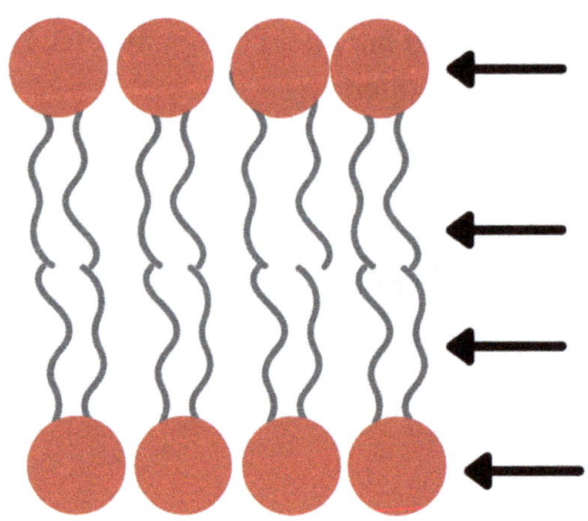

9. Where would you find the lipid bilayer in a cell?

10. Why do phospholipids form a bilayer when they're mixed with water?

Steroids

Steroids have a much different shape than other lipids in that they are made from rings of carbon instead of hydrocarbon chains. Most steroid hormones (testosterone, estrogen, progesterone, etc) are derived from cholesterol, with slight changes made to functional groups to give these hormones different functions. The main function of steroids is to act as messengers to influence cellular processes. Cholesterol is only found in animals: body tissues, cell membranes, and blood. Some examples of steroids are shown in Figure 4.16.

When looking at large organic molecules, C-C bonds are often represented by lines, with the Hydrogens omitted. Wherever there is a bend in the line, you can assume there is a C-C or C=C bond, with the appropriate number of H atoms to complete Carbon four bonds. Functional groups are drawn out.

11. Compare the cholesterol molecule to testosterone and progesterone in Figure 4.16. Circle the structural differences between these molecules.

Figure 4.16. Examples of Steroids

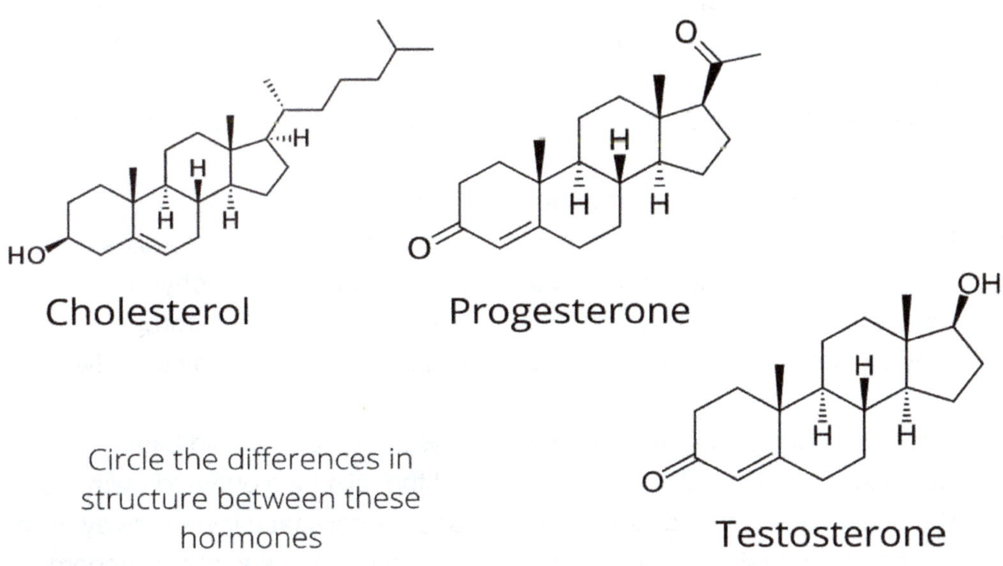

12. Why are steroids classified with other lipids?

13. What is the main function of steroids?

14. From what molecule are steroid hormones derived?

Multiple Choice Review

1. How do saturated and unsaturated fatty acids differ in structure?
    a. Unsaturated fatty acids have double bonds, while saturated fatty acids do not.
    b. Unsaturated fatty acids are straight, while saturated fatty acids are bent.
    c. Saturated fatty acids are found mostly in plants, while unsaturated fatty acids are found mostly in animals.
    d. Unsaturated fatty acids have a carboxyl group, while saturated fatty acids do not.

2. Why are triglycerides considered an efficient way to store energy?
    a. They contain a higher amount of water compared to other molecules.
    b. Each gram of fat contains more calories than a gram of carbohydrate.
    c. Triglycerides can be easily converted into protein for energy storage.
    d. They are composed of short carbon chains that release energy quickly.

3. How do phospholipids arrange themselves in water to form a lipid bilayer?
    a. The heads face inward towards water, and the tails face outward away from water.
    b. The heads face outward towards water, and the tails face inward away from water.
    c. The heads and tails mix randomly in water without any specific arrangement.
    d. The heads and tails form separate layers, with heads on one side and tails on the other.

4. Why do phospholipids have a unique property of being both hydrophilic and hydrophobic?
   a. The phosphate group is hydrophilic, while the fatty acid tails are hydrophobic.
   b. Phospholipids contain equal amounts of water and lipids.
   c. The fatty acid tails are hydrophilic, while the phosphate group is hydrophobic.
   d. Phospholipids have a uniform structure that makes them compatible with water.

5. What distinguishes steroids from other lipids in terms of their molecular structure?
   a. Steroids contain hydrocarbon chains.
   b. Steroids are made from rings of carbon.
   c. Steroids have functional groups attached to their carbon chains.
   d. Steroids are primarily found in plants.

6. How are steroid hormones like testosterone and estrogen derived?
   a. From fatty acids
   b. From amino acids
   c. From cholesterol
   d. From carbohydrates

7. What is the primary function of steroids in biological systems?
   a. To store energy
   b. To act as structural components of cell membranes
   c. To regulate cellular processes as messengers
   d. To provide insulation and protection to organs

8. The lipid molecules we looked at all have different structures and slightly different functions. Why are these molecules all categorized as lipids?
   a. They're all hydrophilic.
   b. They're all hydrophobic.
   c. They are all polar.
   d. They all have the same number of Carbon atoms.

1.1 Introduction to Amino Acids and Proteins

Proteins have many functions in organisms, including:

- Structure: collagen (bone, cartilage, tendon), keratin (hair), actin (muscle)
- Enzymes: amylase, pepsin, catalase, and more than 10,000 others
- Transport: hemoglobin (oxygen)
- Motors: myosin (muscle)
- Hormones: insulin, glucagon
- Receptors: rhodopsin (light receptor in retina)
- Immune system: antibodies

All proteins are made from amino acids. There are 20 different amino acids that differ in the structure of their side chains. For the generic amino acid below, the "R" represents the place holder for the side chain. You should be able to draw the general structure of an amino acid and show how these amino acids join together to produce a protein. There are 20 different Amino Acids but all of them share the same general structure, as seen in Figure 5.1.

Figure 5.1. General Structure of an Amino Acid

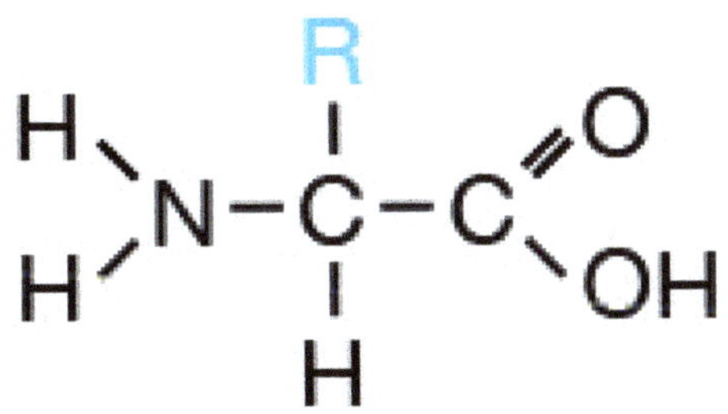

1. Practice drawing the general structure of an amino acid.
   a. At one end there is the NH_2 group. This is referred to as the AMINO group. Draw a circle around it on your diagram and label it.
   b. At the other end there is the COOH group. This is called the CARBOXYL group and is acidic. Draw a circle around it on your diagram and label it.

5

There is an R group sticking up from the central carbon. The R group varies between different amino acids. There are 20 naturally occurring R groups and so there are 20 different amino acids.

You don't need to learn all 20 amino acids for your exam; however you should be able to determine the characteristics, especially hydrophilic or hydrophobic, of a specific amino acid if you are given the R group structure. The table in Figure 5.2 shows the different R groups for each of the 20 Amino acids.

Figure 5.2. Properties of Amino Acids

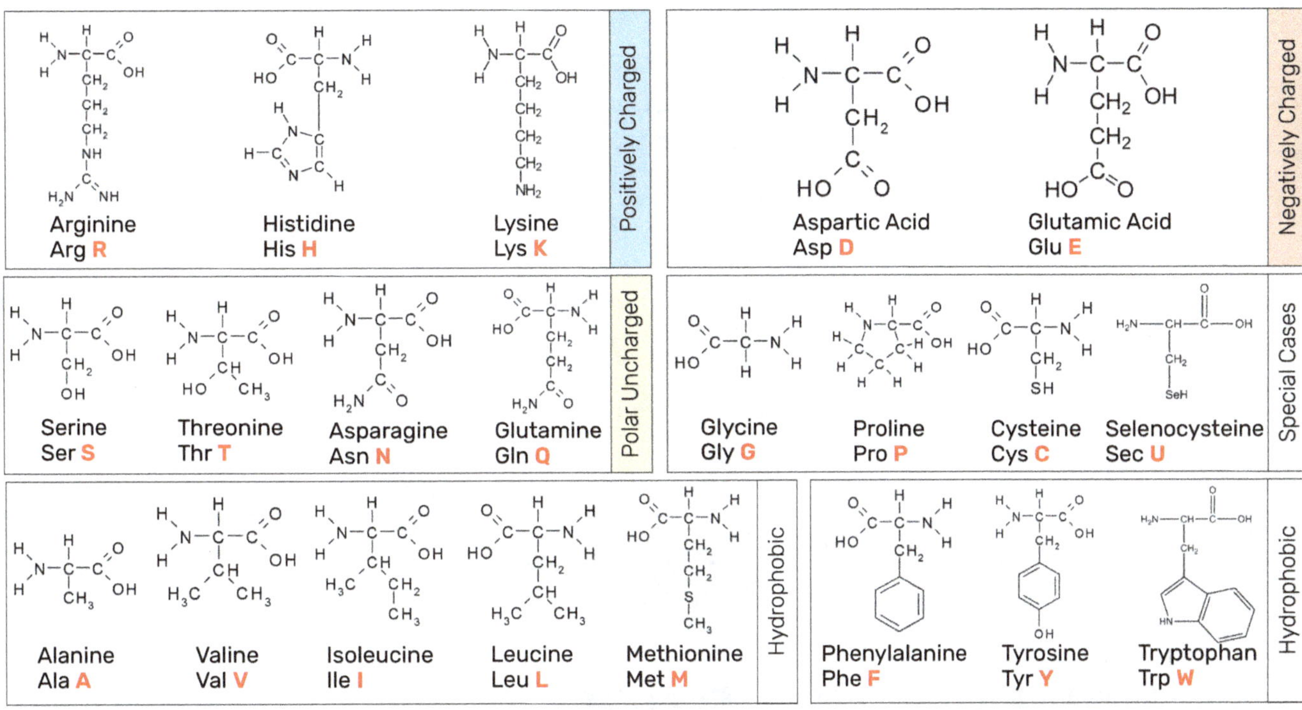

Use Figure 5.2 to help you draw the structure of the following Amino Acids in questions 2–4. Label each one as hydrophilic or hydrophobic based on the characteristics of their side chains.

2. Asparagine

5

3. Leucine

4. Glutamine

Polypeptide Chains

Amino acids can be joined together to form larger molecules by a process called dehydration synthesis. Dehydration synthesis is a chemical reaction in which two molecules are joined together by a covalent bond to form a larger molecule, and at the same time a water molecule is released. This reaction occurs between the functional groups of two separate monomers.

We are going to show how two amino acids can join together through dehydration synthesis. The bond that occurs between amino acids is called a peptide bond and determines the primary structure of a protein. When two amino acids join together, we call it a dipeptide. When many amino acids are joined together a polypeptide is formed. The polypeptide is then folded into its unique three-dimensional shape to form a protein.

5. Draw the synthesis of the peptide bond between two generic amino acids in Figure 5.3.
   a. Draw a circle around where there are two hydrogens and one oxygen close together in Figure 5.3.
   b. Draw a box around the two hydrogen atoms and one oxygen atom that will be removed to release water.
   c. Draw the product (Dipeptide).
   d. Highlight and label the peptide (covalent) bond that is formed.

Figure 5.3. Draw the Synthesis of the Peptide Bond

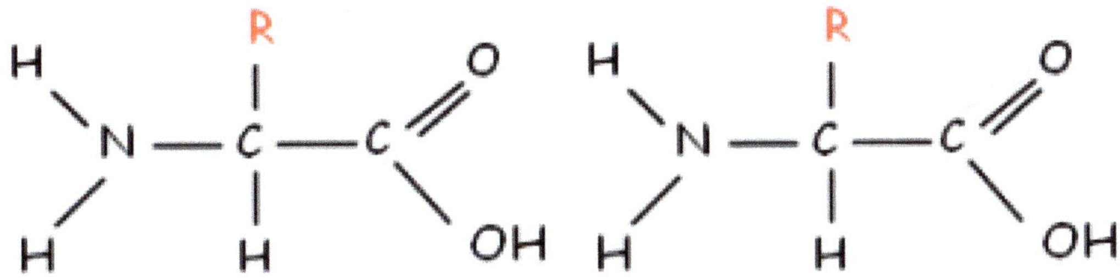

6. Draw and label the products of a dehydration synthesis reaction between the two amino acids valine and methionine. Remember you do not need to learn all 20 amino acids so use the table of amino R groups from earlier in the worksheet to help you.

5

7. Circle or highlight the bond in Figure 5.4. that would need to be broken to separate the tripeptide below into individual amino acids.

Figure 5.4. A Tripeptide

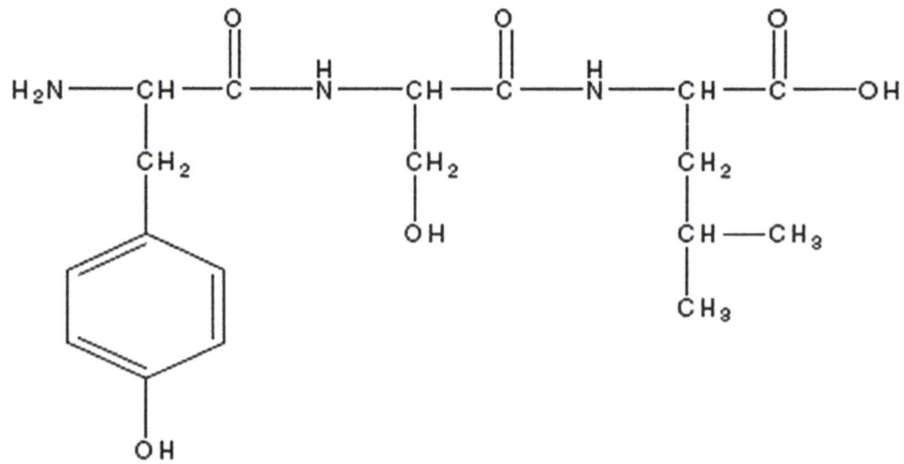

Protein Structure

We already know that a polypeptide is formed by joining many amino acids together (with peptide bonds) to form a long chain. We refer to this chain of amino acids as the primary structure of the protein. Before the protein is ready to perform its function in the body it must be folded. There are several levels of folding that occur before a protein is in the correct shape to carry out its function. These levels of folding are shown in Figure 5.5.

Figure 5.5. The Levels of the Protein Structure

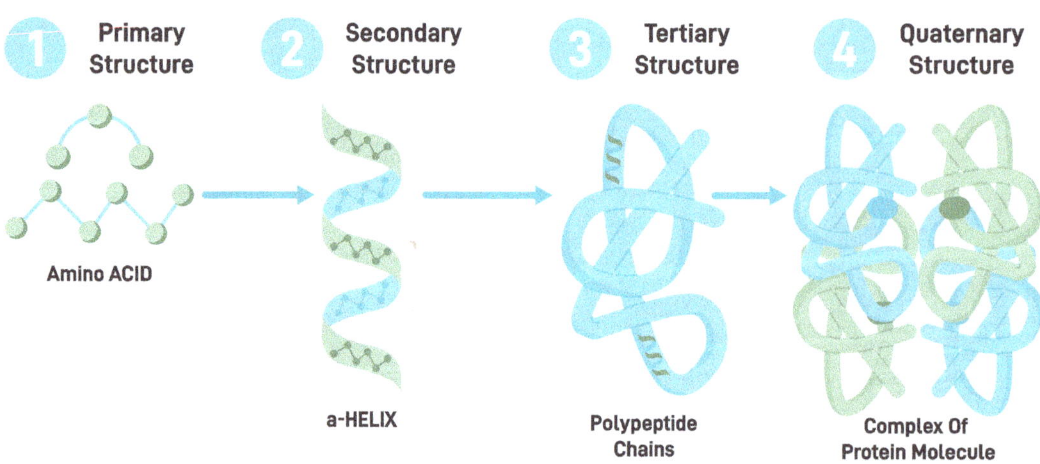

The first folding that occurs to the protein results in a mixture of Alpha-helices and Beta-pleated sheets (you can see what these shapes look like in the diagram below). This first level of folding results in the protein's secondary structure and is a result of hydrogen bonds between the amino and carboxyl groups of the amino acids. The hydrogen bonds responsible for the secondary structure of amino acids is shown in Figure 5.6.

Figure 5.6. Secondary Structure of Proteins

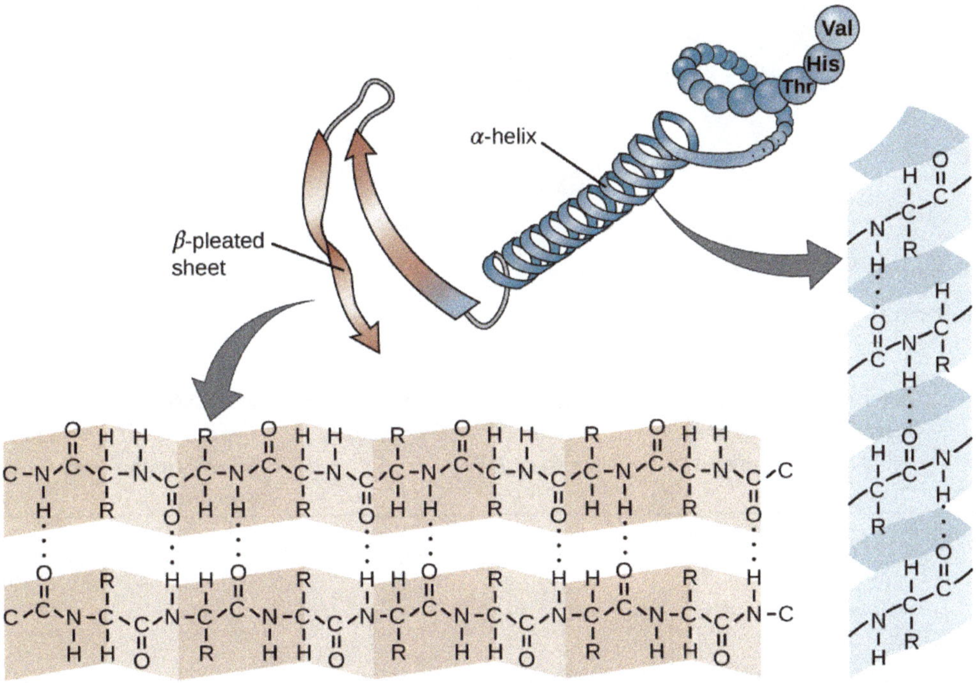

The protein is then folded again on top of its secondary structure to give the protein a very specific three-dimensional (3D) shape. It is this 3D shape that is critical to the protein being able to perform its function. This is called the tertiary structure and is the result of interactions between the side chains of amino acids.

In some proteins, multiple folded peptide chains are combined to form a larger protein. This happens in proteins such as hemoglobin which is used to transport oxygen. The final hemoglobin protein has four peptide chains that combine to form the final protein structure. This is called the quaternary structure.

There are different types of bonds responsible for providing the protein with its levels of structures. The weakest and most easily broken bonds are Hydrogen bonds. Ionic bonds are stronger, but they can be broken by changes in pH. The strongest bonds in the protein (and hardest to break) are the covalent bonds. There are two types of covalent bonds involved in protein structure. These are the peptide bonds and the disulphide bridges. The types of bonds responsible for the tertiary structure of proteins is shown in Figure 5.7.

Figure 5.7. Tertiary Structure of Proteins

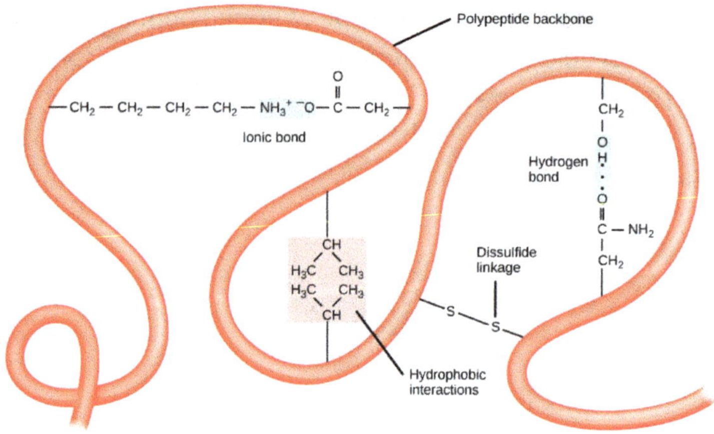

8. Use the information in this section to complete Table 5.1.

Table 5.1.

Level	Definition/ Description	Peptide bonds present?	Hydrogen bonds present?	Ionic bonds present?	Hydrophobic/ hyedrophilic interactions present?	Disulfide bridges present?
Primary						
Secondary						
Tertiary						

Multiple Choice Review

1. What name is given to the part of the amino acid molecule that varies between one type of amino acid and the next?
   a. P group
   b. R group
   c. Q group

2. What is the name of the bond that forms between two amino acids when they are joined by a dehydration synthesis reaction?
   a. Peptide
   b. Glycosidic
   c. Amine

3. Which of these bonds is hardest to break?
   a. Peptide
   b. Ionic
   c. Hydrogen

4. Which of these bonds is only present in the protein's tertiary structure?
   a. Peptide
   b. Ionic
   c. Hydrogen

5. What is the name given to the molecule formed when two amino acids join together?
   a. Protein
   b. Diamine
   c. Dipeptide

6. Which molecule is always produced in a dehydration synthesis reaction?
   a. Dipeptide
   b. Water
   c. Amino Acid

1.2 Enzymes

Cells are constantly carrying out chemical reactions. These reactions perform functions such as digestion, respiration (so you have energy), sending signals and creating new cell parts. Everything that your body does, and the maintenance of homeostasis, is achieved through chemical reactions. However, these chemical reactions would not happen quickly enough to maintain life without enzymes.

Enzymes are biological catalysts which means they speed up reactions in the body. Enzymes are proteins which are made in such a way that reacting molecules fit into them at a location called the active site. If you look at Figure 5.8, you can see the substrate fits perfectly into the active site (the part of the enzyme that does the reactions). We call this the lock and key model because the enzyme fits in the substrate like a lock and key.

Figure 5.8. General Diagram of Enzyme Function

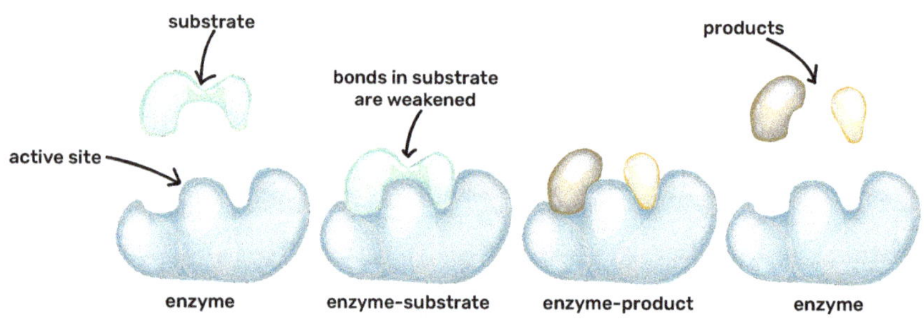

Enzymes can break down substrates, as in the image to the right (this typically happens in digestion and respiration) or they can put two substrates together (this happens when our cells build proteins using amino acids from the foods we eat). But it is important to remember that generally one enzyme can only react with one substrate.

Figure 5.9. The Effect of Enzymes on Activation Energy of a Reaction

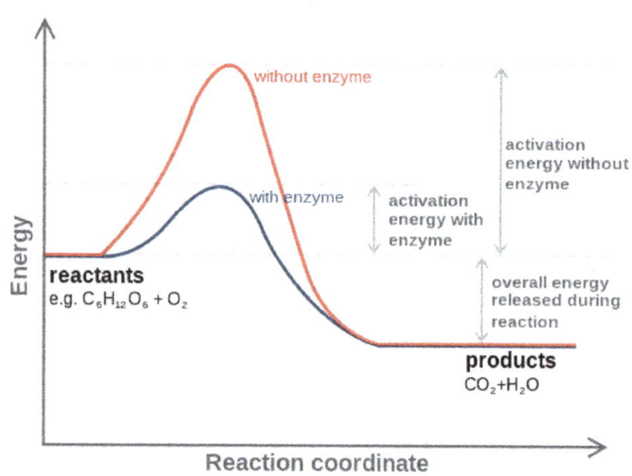

When the enzyme is bonded to its substrate, this strains the bonds within the molecule to promote the formation of new or different bonds, or potentially strains the bonds so they break more easily. By straining the bonds in the substrate, the enzyme lowers the activation energy necessary for a chemical reaction to occur, as shown in Figure 5.9. The lower the activation energy is, the faster the reaction happens.

1. Enzymes are "catalysts." What does that word mean?

5

2. Label Figure 5.10 with the following words:
   - Substrate
   - Active site
   - Product
   - Enzyme

Figure 5.10. Diagram of Enzyme Function

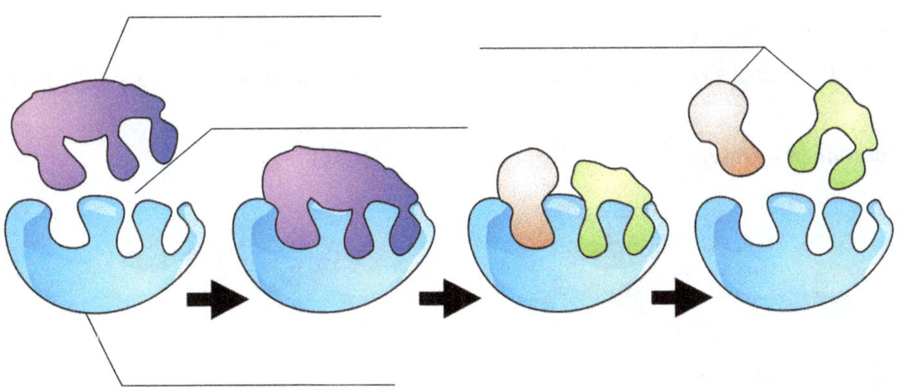

3. Is the enzyme in Figure 5.10 performing dehydration synthesis or hydrolysis? How do you know?

4. Explain why enzymes are said to be specific. What physical part of the enzyme determines this specificity?

Enzymes, like other proteins, are sensitive to their environment. When placed in a pH or temperature outside their ideal range, their structure changes significantly, and they become denatured. This means that enzymes are specific to their environment, for example, the enzymes in your stomach are suited to work at a pH of about 2 because your stomach is acidic. If you put them into a different pH, they will become denatured (they won't work). Enzymes lose their function when they are denatured because their active site will change shape. Because the active site has changed shape, the substrate will no longer fit, and the enzyme won't be able to catalyze reactions. Just so you know, enzymes do not die because they were never alive in the first place, so they can't die. They only denature. Depending on the extent of denaturation, the enzyme may be refolded, or it will be digested by other enzymes to recycle the amino acids to be used in other proteins. A diagram of a protein denaturing is shown in Figure 5.11.

Figure 5.11. Denaturation of a Protein

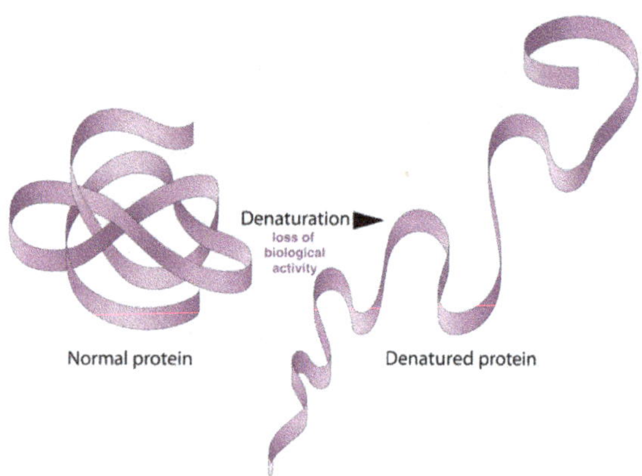

5. What is Energy of Activation (Ea)? How do enzymes relate to this?

6. Consider the following enzymatic pathway where enzymes convert substance A into others.

   A → C → D
   ↓
   B

   If the enzyme responsible for converting A into C were not working, what would happen?
   a. A levels would increase; B, C, and D levels would decrease.
   b. A and B levels would increase; C and D levels would decrease.
   c. A, B, and C levels would increase; D levels would decrease.
   d. A, B, C, and D levels would all decrease.

7. What might cause an enzyme to not work?

5

Enzymes will sometimes bond to other molecules that are not proteins as part of their structure. A molecule is considered an enzyme activator if it increases the reaction rate of the enzyme. Usually, enzyme activators bond to an allosteric site, meaning the activator bonds somewhere else on the enzyme. When this happens, the activator usually increases the affinity of the enzyme's active site for its substrate. Molecules that activate enzymes are typically called cofactors or coenzymes. A diagram of how cofactor or coenzymes work is shown in Figure 5.12.

Figure 5.12. Enzyme Cofactor and Coenzymes

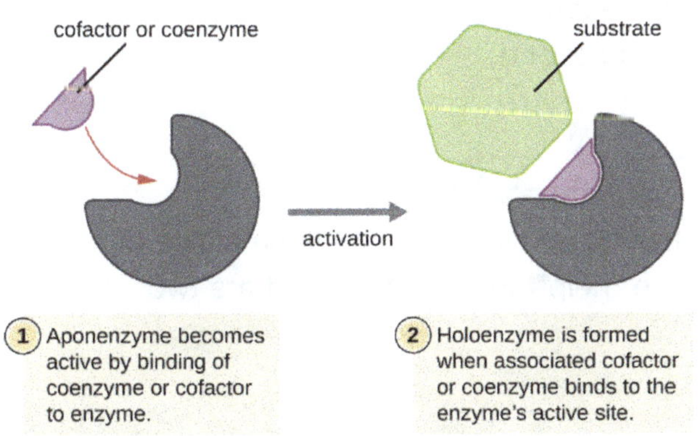

Some molecules, when bonded to an enzyme, will inhibit, or prevent, the enzyme from functioning, as shown in Figure 5.13. Inhibition can be permanent or temporary. Inhibitors can bond to the active site, and therefore block the active site from bonding to the substrate. This is called competitive inhibition since the inhibitor is "competing" with the substrate for the active site. Or, inhibitors can bond to an allosteric site, which usually causes the enzyme to change shape, which also alters the shape of the active site and prevents it from bonding with the substrate. This is called non-competitive inhibition. You've likely taken enzyme inhibitors when you've had a headache or been in pain. Non-steroid anti-inflammatory drugs (NSAIDS) like aspirin and ibuprofen are enzyme inhibitors. They inhibit enzymes that produce prostaglandins, molecules that trigger inflammation and pain when tissue is damaged.

Figure 5.13. Enzyme Inhibition

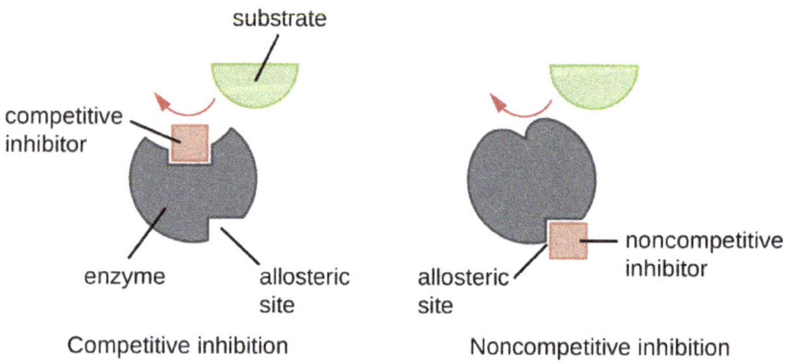

8. Consider the following enzymatic pathway where enzymes convert substance A into others.

   $A \rightarrow C \rightarrow D$

   $\downarrow$

   B

   The enzyme for A into C works at 10 molecules per second. The enzyme for C into D works at 5 molecules per second. Ignoring molecule B, what will happen to the level of D?

   a. D will be made at 10 molecules per second.
   b. D will be made at 5 molecules per second.
   c. D will be made at 7.5 molecules per second.
   d. D will not be made at all.

9. Suppose the faster enzyme from the question above is due to the presence of an "enzyme helper." How can a substance "help" an enzyme? What are two examples of these helpers?

10. Let's say we want to develop a compound that would prevent product D from being produced for a short period of time. What type of molecule are we trying to create? What enzyme pathway would we want this molecule to target?

5

Multiple Choice Review

1. What is the primary function of enzymes in the body?
   a. To provide energy directly to cells
   b. To act as biological catalysts that speed up chemical reactions
   c. To serve as structural components of cell membranes
   d. To store genetic information

2. What typically happens when an enzyme binds to its substrate?
   a. The substrate is expelled from the cell.
   b. The enzyme permanently changes shape.
   c. The bonds within the substrate are strained, facilitating a reaction.
   d. The enzyme and substrate form a permanent bond.

3. How do enzymes affect the activation energy of a chemical reaction?
   a. They increase the activation energy required.
   b. They decrease the activation energy required.
   c. They have no effect on the activation energy.
   d. They variably affect the activation energy depending on the reaction type.

4. Which statement best describes the specificity of enzymes?
   a. Each enzyme can react with any substrate.
   b. Enzymes can react with a range of similar substrates.
   c. One enzyme generally reacts with one specific substrate.
   d. Enzymes select substrates based on the size of the substrate molecule.

5. What happens to enzymes when they are exposed to a pH level outside their ideal range?
   a. They become more efficient.
   b. They become denatured.
   c. They increase the rate of reaction.
   d. They change the pH of the environment.

6. Why do enzymes lose their function when denatured?
   a. The substrate becomes unstable.
   b. The enzyme produces more substrates.
   c. The active site changes shape, preventing the substrate from fitting.
   d. The enzyme absorbs extra hydrogen ions.

7. What is the fate of denatured enzymes that cannot be refolded?
   a. They remain unchanged in the system forever.
   b. They are digested by other enzymes to recycle their amino acids.
   c. They are expelled from the body as waste.
   d. They are stored in the cell for future use.

8. How do non-steroidal anti-inflammatory drugs (NSAIDs) like aspirin and ibuprofen function as enzyme inhibitors?
   a. They increase the production of prostaglandins.
   b. They inhibit enzymes that produce prostaglandins.
   c. They act as cofactors to enhance enzyme activity.
   d. They serve as substrates for pain-triggering enzymes.

9. What is the role of an enzyme activator?
   a. To decrease the reaction rate of the enzyme
   b. To increase the reaction rate of the enzyme
   c. To permanently deactivate the enzyme
   d. To serve as a primary substrate for the enzyme

6.1 Cell Membrane Structure and Function

Phospholipids

Phospholipids are similar to triglycerides except that a phosphate group replaces one of the fatty acids. The phosphate group is a very polar molecule. This gives the phospholipid the unique property of one portion of the molecule being hydrophilic (the phosphate group) and the rest of the molecule being hydrophobic (the fatty acids). Phospholipids are the main component of cell membranes.

1. Fill in one phosphate group and the remaining fatty acid. Usually, one fatty acid in phospholipids is unsaturated. Label all the parts in Figure 6.1.

Figure 6.1. Draw in the Phospholipid Structure

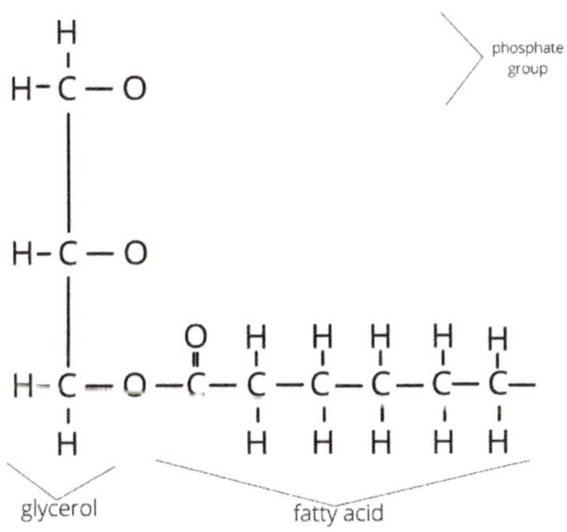

Phospholipids are commonly drawn as a circle with two tails. Phospholipids are polar on the phosphate group side and nonpolar on the fatty acid side. When phospholipids are mixed with water, they line up with heads on one side facing out towards water and their tails on the inside, away from water. One layer of phospholipids is called a micelle. Two rows of phospholipids are called a lipid bilayer, and they make up the main component of all cell membranes, as shown in Figure 6.2.

Figure 6.2. Phospholipids

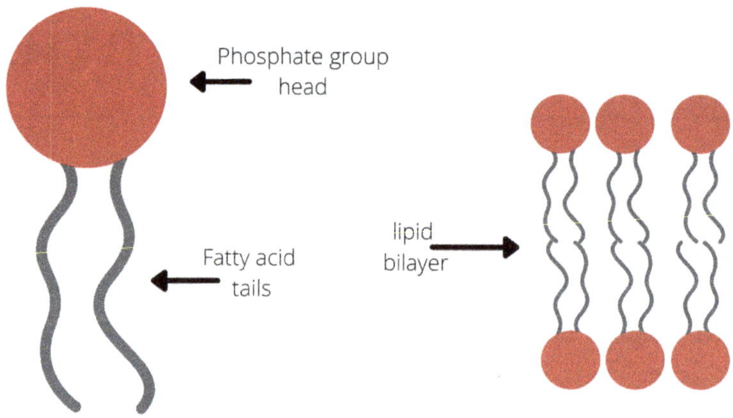

2. Label the regions in Figure 6.3 in the bilayer below as polar or nonpolar.

Figure 6.3. Phospholipid Bilayer

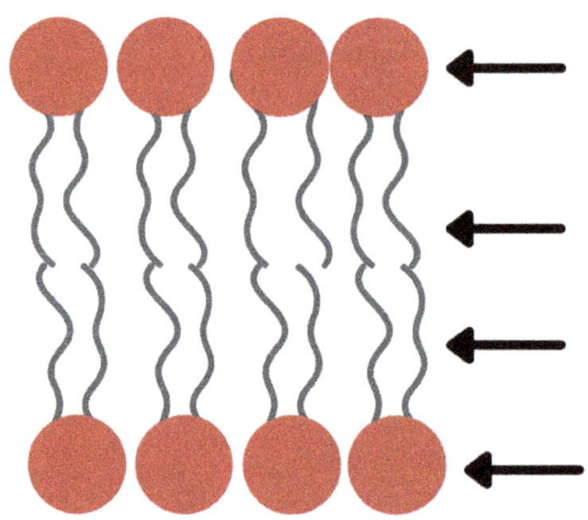

3. Where would you find the lipid bilayer in a cell?

4. Why do phospholipids form a bilayer when they're mixed with water?

5. Why would it be important for one of the fatty acid tails to be unsaturated? (hint: read the paragraph below)

6

Structure of the Membrane

Fluid Mosaic Model

Our current understanding of the cell membrane is that it is mainly composed of phospholipids with embedded proteins, as shown in Figure 6.4. The phospholipid bilayer is fluid, and the proteins are embedded into the bilayer like a mosaic. The proteins that are embedded in the membrane have a variety of functions, as we'll explore later. Each type of specialized cell has its own set of membrane proteins that help enable its function, and the type of proteins in the membrane can change depending on the environment the cell is in or the signals it receives from other cells.

Additionally, the cell membrane has cholesterol molecules within the hydrophobic layer that help maintain the integrity of the membrane in extreme temperatures.

Figure 6.4. Fluid Mosaic Model of the Cell Membrane

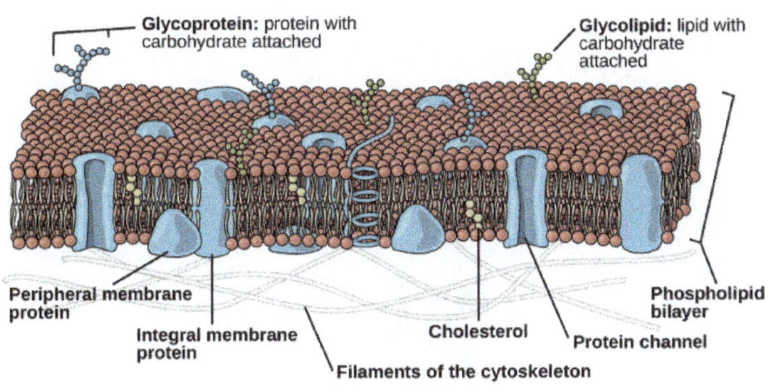

6. Why do the phosphate heads of the phospholipids face the cytoplasm and the extracellular fluid?

7. Explain what the term “fluid mosaic” means.

6

8. What molecules maintain the membrane’s fluidity?

Types of Membrane Proteins

Integral membrane proteins are permanently embedded in the phospholipid bilayer. These can be either transmembrane proteins (meaning they span to both sides of the membrane) or monotypic (meaning they are only on one side of the membrane, the cytoplasm side or the extracellular side). These proteins play a major role in transporting materials across the membrane attaching cells to other cells, and they can act as molecular markers for cellular communication. Common examples of integral membrane proteins are ion channels that allow ions like Na^+, K^+, Ca_2^+, Cl^-, and many others to pass through, like what’s shown in Figure 6.5.

Figure 6.5. Channel Protein That Can Be Regulated

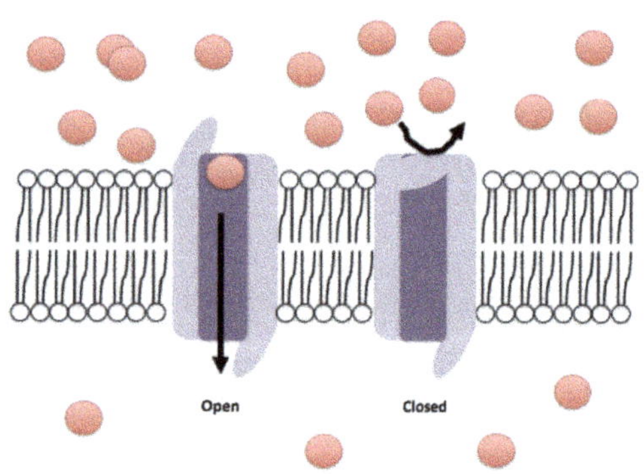

Peripheral membrane proteins are proteins that are added or removed from the membrane relatively easily and are usually temporary. These will often attach to integral proteins. Peripheral membrane proteins often function in the regulation of integral proteins, attachment to the cytoskeleton or extracellular matrix, communication through signal transduction, and as enzymes.

9. Describe the roles of proteins embedded in the cell membrane.

Cell Transport

But why do cells need proteins to help transport molecules across the membrane? Because the membrane is **selectively permeable** or **semi-permeable**. This means that some molecules can pass directly through the phospholipid bilayer, and others cannot. This also means that cells can regulate what molecules they allow to pass through.

So, what can pass through the phospholipid bilayer? Small, non-charged, and non-polar molecules. Things like O_2 and CO_2 can easily pass through the phospholipid bilayer and will move to areas of lower concentration, as shown in Figure 6.6.

Figure 6.6. Diffusion Across the Membrane

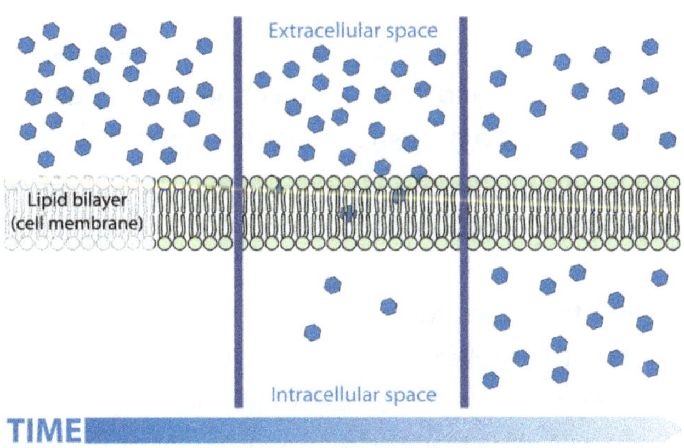

Larger molecules, or anything that has a charge (like an ion) or is polar moves across the membrane through transport proteins. **Transport proteins** can be channel proteins, like pores through the membrane. They can also be carrier proteins that bind to a specific solute and change shape to move it across the membrane, as shown in Figure 6.7. When molecules move across the membrane through diffusion, they will move to the side with a lower concentration until there is an equal concentration of the molecule on either side of the membrane.

Figure 6.7. Transport Proteins

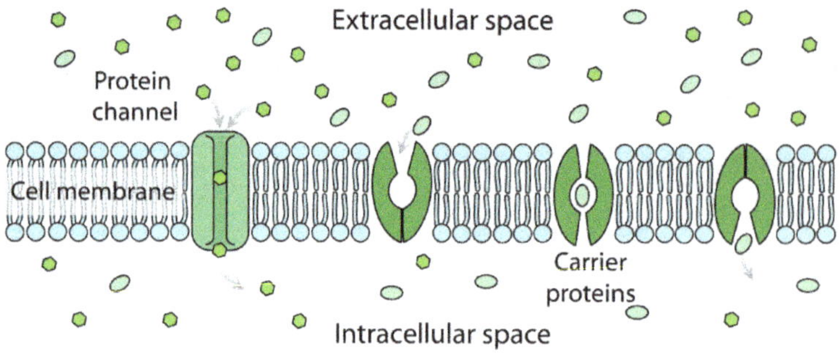

10. Why would small, non-charged molecules be able to pass directly through the phospholipid bilayer? Think about the chemical properties of the bilayer.

11. Why do ions require transport proteins to move across the membrane?

Transport of molecules across the membrane can be categorized as passive or active. Passive transport does not require the cell to expend energy (ATP) and is the result of diffusion. Passive transport can occur with or without a transport protein. If a protein is involved in passive transport, it is called **facilitated transport**. Osmosis, water moving across the membrane to an area with a higher concentration of solute, is a type of passive transport.

Active transport is a type of membrane transport that does require the cell to expend energy. Cells expend energy to transport molecules when the molecules are moving in the opposite direction that they would naturally diffuse, or in other words, when the molecules are moving from a low concentration to a high concentration. Active transport is an important process that allows cells to maintain homeostasis despite changing external environments. The epithelial cells that line our intestines use active transport to bring food molecules into their cytoplasm, moving these molecules from an area of low concentration (the intestinal tract) to an area of high concentration (into the intestinal epithelial cell).

Another type of active transport is when cells use vesicles to move large amounts of substances into or out of the cell in a process known as **exocytosis** (moving outside the cell) and

endocytosis (moving inside the cell). This is considered active transport because the has to expend energy to move the membrane and vesicles around. Endocytosis is usually used to bring in very large volumes of materials into the cell, like when a large cell eats a smaller cell. This cell eating a cell is a type of endocytosis called **phagocytosis**, as seen in Figure 6.8. Exocytosis is used to send out large volumes of molecules, such as hormones or waste, in a short period of time.

Figure 6.8. Phagocytosis and Exocytosis

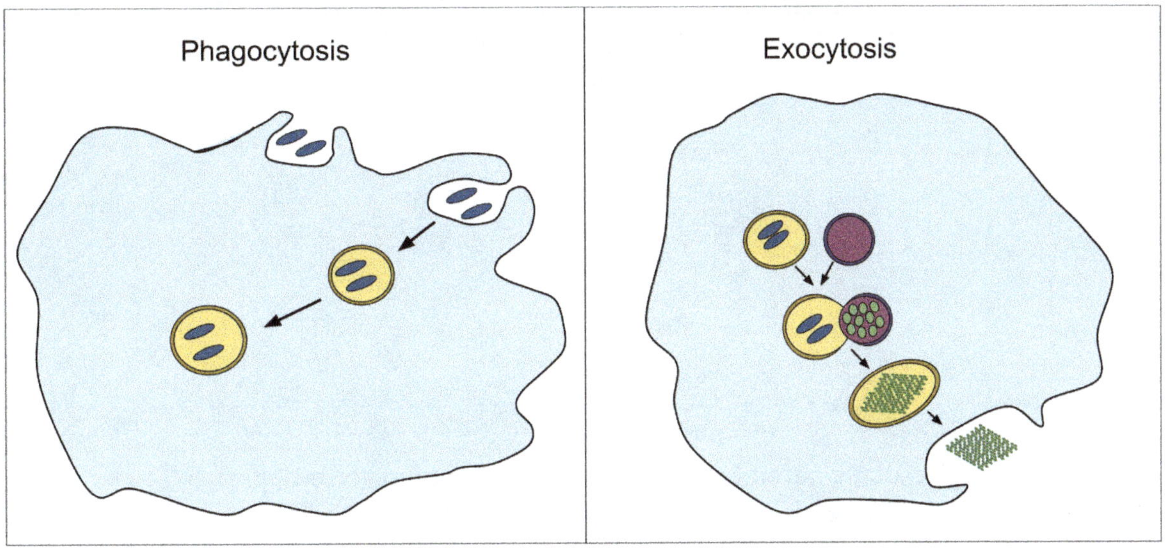

12. What is the difference between diffusion and facilitated transport?

13. How is active transport different from passive transport?

14. In what types of scenarios do cells use active transport?

15. Figure 6.9 features several snapshots of molecules being transported into a cell via a transport protein. What type of transport do you think this is? Explain your reasoning.

Figure 6.9.

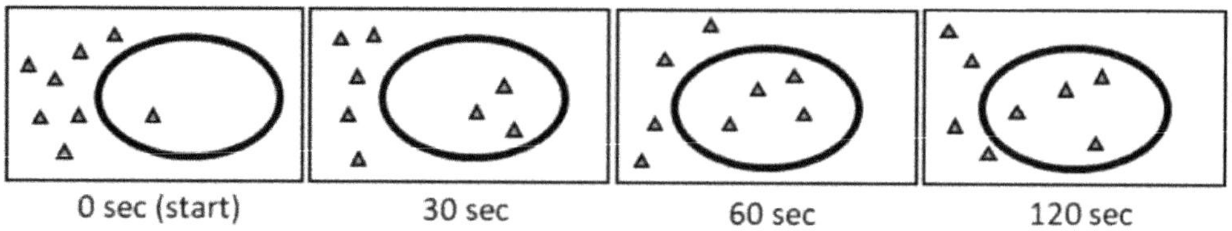

16. Figure 6.10 features several snapshots of a different molecule being transported into a cell via a transport protein. What type of transport do you think this is? Explain your reasoning.

Figure 6.10.

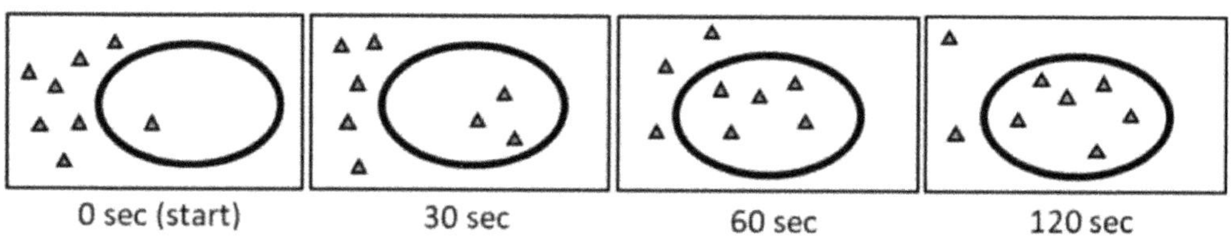

Osmosis

Osmosis is a type of diffusion. Specifically, **osmosis** is the diffusion of water molecules from a high concentration to a low concentration. However, it gets tricky because when talking about solutions and osmosis, we don't use the concentration of water to discuss osmosis. Instead, we use the concentration of solute to determine where water will move. Remember that a solution is solute (what gets dissolved) and solvent (what does the dissolving, usually water for living things). So, if we have a 5% glucose solution, that means we have 95% water and 5% glucose.

The number one rule of osmosis is that water will move across the membrane to the side with more solute. When discussing osmosis, we use tonicity to describe the relative concentration of solutes. These terms should only be used to compare two solutions. The direction of net water movement in different tonicities is shown in Figure 6.11 and explained further in Table 6.1.

6

Figure 6.11. Osmosis in a Red Blood Cell

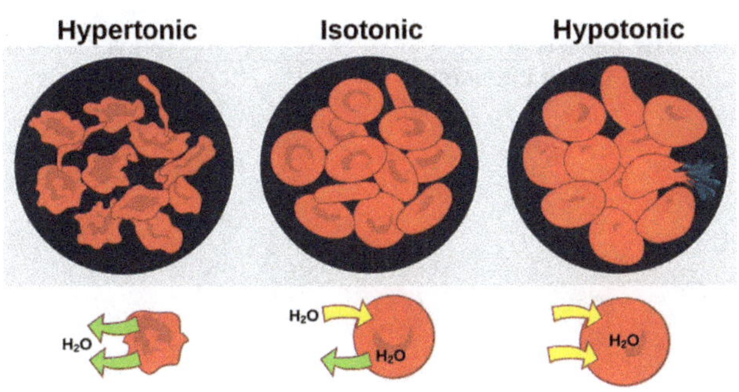

Table 6.1. Direction of Water Movement in Osmosis

Tonicity	Example	Direction of water movement
Isotonic: two solutions have the same concentration of solutes	A cell (with 0.9% solute concentration in its cytoplasm) is placed in a 0.9% NaCl solution	Water is moving both into and out of the cell in equal amounts
Hypotonic: a solution has a lower concentration of solute compared to another solution	A cell (0.9% solute) is placed in a solution of pure water	Water moves into the cell as there is a greater concentration of solute on that side of the membrane
Hypertonic: a solution has a higher concentration of solute compared to another solution	A cell (0.9% solute) is placed in a 10% NaCl solution	Water moves out of the cell as there is a greater concentration of solute on that side of the membrane

17. The diagrams below show the concentration of water and salt inside the cell (in the circle) and the concentration of water and salt surrounding the cell. Complete the sentence by comparing the concentration of the water inside the cell and the concentration outside the cell.

1.

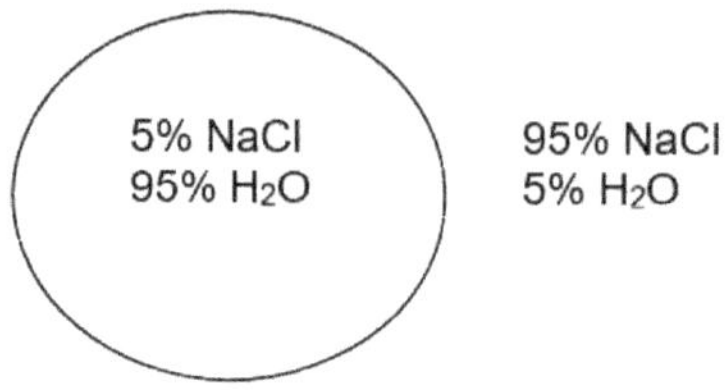

a. Water will flow ______________________ (into the cell, out of the cell, in both directions).

b. The cell will ______________________ (shrink, burst, stay the same).

2.

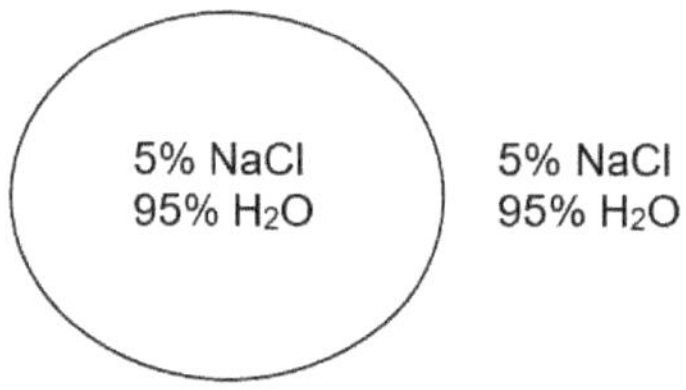

a. Water will flow ______________________ (into the cell, out of the cell, in both directions).

b. The cell will ______________________ (shrink, burst, stay the same).

3.

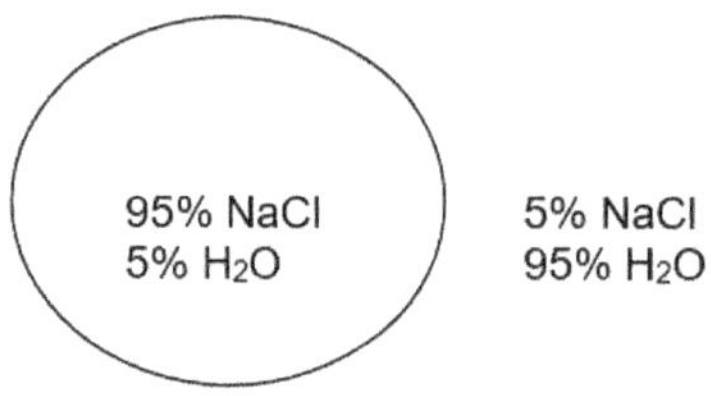

a. Water will flow ______________________ (into the cell, out of the cell, in both directions).

b. The cell will ______________________ (shrink, burst, stay the same).

18. Most sports drinks are isotonic in relation to human body fluids. Explain why athletes should drink solutions that are isotonic to body fluids when they exercise rather than ones that are hypertonic to body fluids (contain a greater proportion of solute in comparison to the fluids in and around human body cells).

19. Examine Figures 6.12–6.16 and identify each image as: diffusion, osmosis, facilitated diffusion, active transport, exocytosis, or endocytosis.

Figure 6.12.

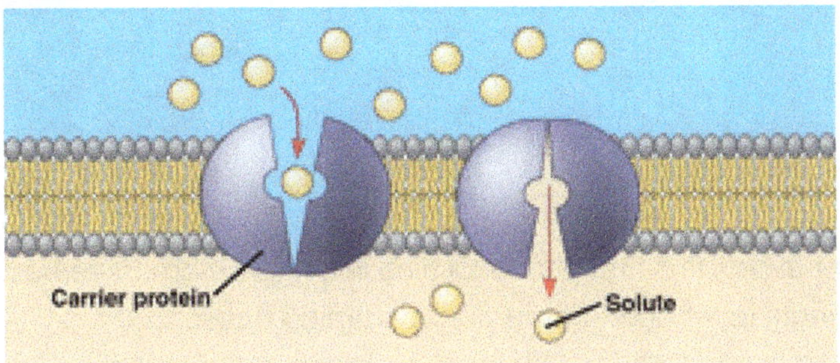

Figure 6.13.

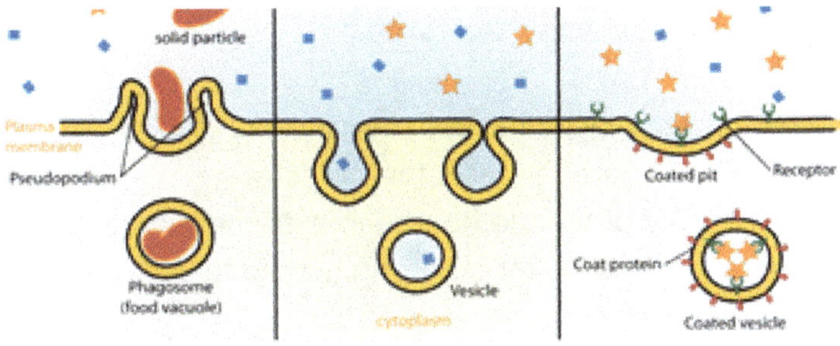

Figure 6.14.

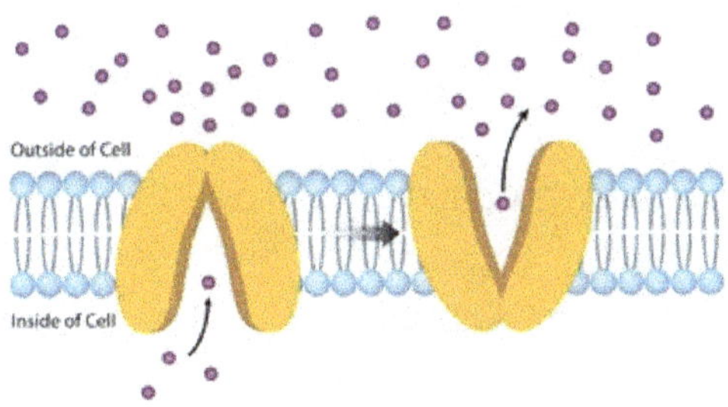

Figure 6.15.

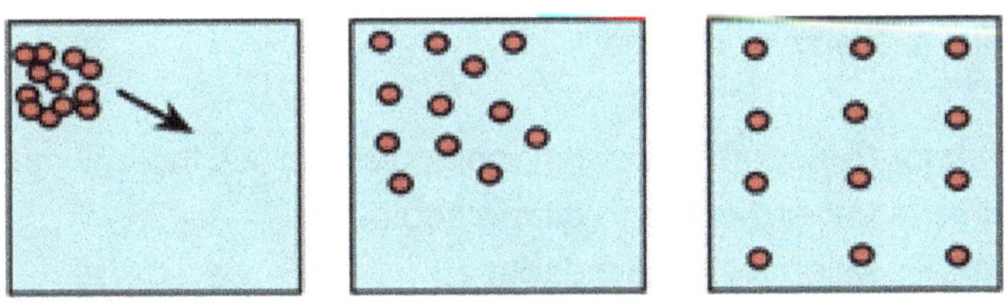

Figure 6.16.

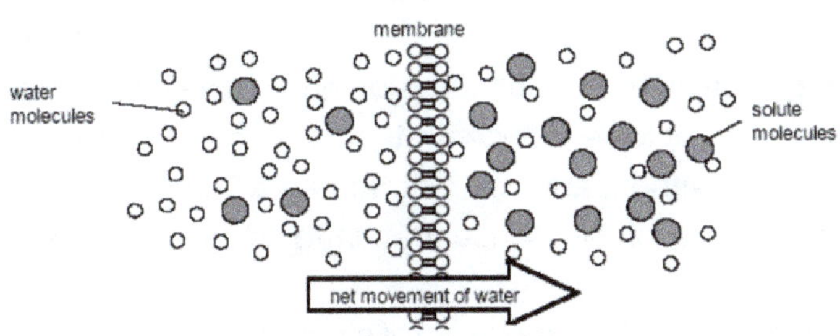

Multiple Choice Review

1. What is the primary structural component of the cell membrane?
    a. Proteins
    b. Cholesterol
    c. Phospholipids
    d. Carbohydrates

2. How is the arrangement of proteins and phospholipids in the cell membrane best described?
    a. Proteins form a single layer outside the phospholipid bilayer.
    b. Proteins are randomly scattered within a solid phospholipid layer.
    c. A fluid mosaic of phospholipids with embedded proteins.
    d. A rigid structure of phospholipids with fixed proteins.

3. What role do cholesterol molecules play in the cell membrane?
    a. They primarily function as transport channels for molecules.
    b. They help maintain the integrity of the membrane in extreme temperatures.
    c. They act as the primary receptor sites for signaling molecules.
    d. They provide energy for membrane repair.

4. Why might the proteins in a cell membrane change?
    a. Due to changes in the genetic makeup of the cell.
    b. In response to environmental changes or signals received from other cells.
    c. As a result of aging and wear on the cell.
    d. They change daily as part of normal cellular processes.

5. Which type of molecules can easily pass through the phospholipid bilayer without help?
    a. Large, polar molecules
    b. Charged ions
    c. Small, non-charged, non-polar molecules
    d. All molecules, regardless of size or charge

6. Why are proteins necessary for transporting certain molecules across the cell membrane?
    a. Because the membrane is impermeable to all molecules
    b. Because the membrane is selectively permeable
    c. Because proteins provide the energy needed for transport
    d. Because all molecules can pass freely without assistance

7. How does the movement of molecules through diffusion across the cell membrane occur?
    a. Molecules move from areas of lower concentration to areas of higher concentration.
    b. Molecules move randomly without any concentration gradient.
    c. Molecules move from areas of higher concentration to areas of lower concentration until balanced.
    d. Molecules require active transport to move across the membrane.

8. If a cell is placed in a hypotonic solution, what will likely happen to the cell, and why?
   a. The cell will shrink because water moves out of the cell.
   b. The cell will swell because water moves into the cell.
   c. The cell will remain unchanged because the water movement is balanced.
   d. The cell will burst because it loses all its water.

9. In medical treatment, why is it important to administer isotonic solutions intravenously instead of pure water?
   a. To prevent cells from swelling and bursting due to excess water intake
   b. To encourage cells to expel toxins by shrinking
   c. To increase the osmotic pressure within blood vessels
   d. To decrease the osmotic pressure within blood vessels

10. How would plants react if watered with a very salty solution, considering the principles of osmosis?
   a. They would thrive as the salty solution is beneficial.
   b. They would wilt because water is drawn out of their root cells.
   c. They would grow rapidly due to increased nutrient uptake.
   d. They would change color due to chemical reactions with salt.

11. How would a cell compensate if a vital ion needed for cellular function is in lower concentration inside the cell than outside?
   a. The cell would use passive transport to balance ion concentrations.
   b. The cell would employ active transport to pump ions against the concentration gradient.
   c. The cell would stop all transport processes to conserve energy.
   d. The cell would increase the permeability of the membrane to ions.

7.1 Cells

A cell is the basic unit of life, the smallest structural and functional unit capable of performing all the necessary processes of life. Each cell is a self-contained living entity surrounded by a plasma membrane that can metabolize nutrients, replicate, grow, respond to stimuli, and, in multicellular organisms, contribute to the larger tissue or organ system of which it is composed. Cells vary widely in size and shape and thus function. They can exist as independent organisms, such as bacteria and yeast, or can form part of larger organisms, like the millions of cells in plants and animals. In more complex organisms, cells differentiate and specialize, taking on specific structures, allowing them to play specific roles and contribute to the organism's overall functioning and survival. This division of labor among specialized cells allows multicellular organisms to perform more complex and adaptive functions. Cells are typically composed of a membrane that encloses cytoplasm, genetic material (DNA), and various organelles, each fulfilling distinct roles necessary for maintaining homeostasis and carrying out the cell's functions.

1. Why are cells considered the smallest unit of life?

Eukaryotic and Prokaryotic Cells

Cells are broadly categorized into two main types: prokaryotic and eukaryotic. These two types of cells are diagrammed in Figure 7.1. **Prokaryotic cells**, the cell type of bacteria, are simpler and smaller in structure because they do not possess many of the organelles seen in eukaryotic cells. Prokaryotes do not have a defined nucleus and other membrane-bound organelles. Instead, their DNA floats freely within the cytoplasm of the cell in a region called the nucleoid. **Eukaryotic cells**, which make up plants, animals, fungi, and a diverse group of organisms called protists, are generally larger and more complex. They possess a distinct nucleus, a membrane-bound organelle that houses their DNA. Eukaryotes also have various specialized organelles like mitochondria and the endoplasmic reticulum, each composed of specialized membrane-bound compartments. This organization allows eukaryotic cells to perform more advanced and specialized functions, facilitating greater structural and functional diversity within organisms. We will spend most of our time on Eukaryotic cells and their unique structures.

2. What are the main differences between prokaryotic and eukaryotic cells?

Figure 7.1. Diagram of the Prokaryotic and Eukaryotic Cells

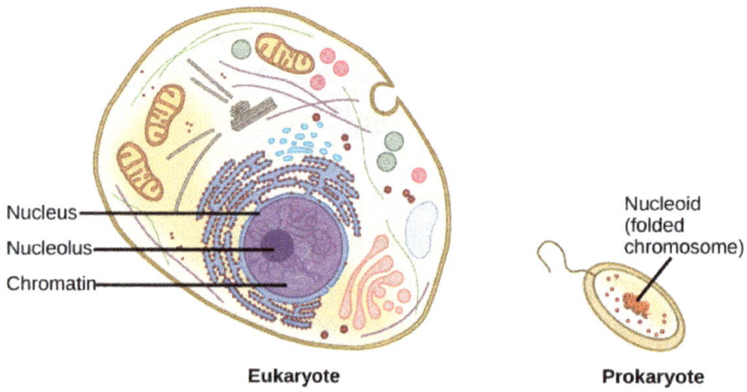

Eukaryotic Cell Structures

Animal cells, the basic units of life in multicellular organisms, are complex structures surrounded by a plasma membrane and filled with various organelles that each perform specific functions essential for the cell's and ultimately the organism's survival. Directing the functions of the cell is the nucleus, which acts as the control center by housing DNA and regulating all cellular activities including protein synthesis and cell division, crucial for maintaining cellular homeostasis.

The Endomembrane System

Surrounding the nucleus, we find the cell's manufacturing and packaging sites, which include the endoplasmic reticulum (ER) and the Golgi apparatus, as seen in Figure 7.2. The nucleus, the ER, and the Golgi are collectively called the endomembrane system and function together in producing molecules. These organelles, like many others, are specialized compartments of folded-up phospholipid bilayer. Because most organelles are composed of phospholipid bilayer, this allows parts of organelles to be exchanged with other organelles, which we'll explore as we investigate how the endomembrane system works. Let's start out at the rough ER. The rough ER is important for the synthesis and folding of proteins.

This part of the ER appears bumpy due to the many ribosomes embedded into it. Ribosomes are the organelles that catalyze the bonds between amino acids to build proteins. The smooth ER (without ribosomes) is involved in the production of lipids and the breakdown of toxic molecules. Once a molecule is built at the ER, for example, an enzyme like amylase used to break down starch, it is packaged in a bubble composed of phospholipids called a vesicle for transport to the Golgi. The Golgi apparatus modifies, sorts, and packages proteins and lipids for transport out of the cell, ensuring that substances necessary for cell structure and function are correctly processed and distributed. But how does the vesicle travel to the Golgi? This is where the cytoskeleton comes in.

Figure 7.2. The Endomembrane System

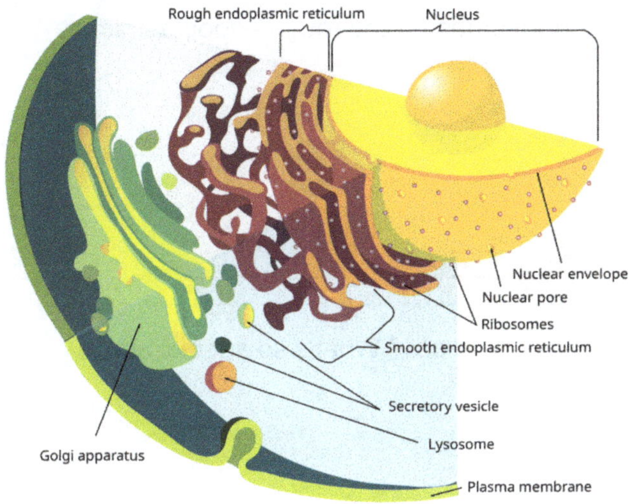

The cytoskeleton, as seen in Figure 7.3 is a network of proteins throughout the cytoplasm of the cell that are used to anchor organelles in place, providing structural support for the cell. The cytoskeleton also provides a highway system for the movement of material, like vesicles. So, our amylase protein was built by ribosomes at the rough ER, was transported to the Golgi via a vesicle that was pinched off from the rough ER, and then the vesicle fused with the Golgi apparatus, allowing enzymes to modify the newly constructed enzyme into its final functional form. This enzyme will then be shipped off to its final destination in another vesicle.

Figure 7.3. The Cytoskeleton

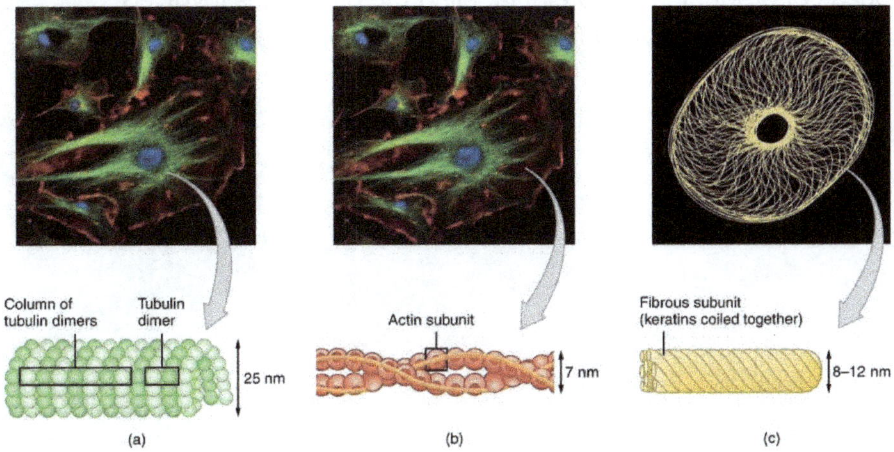

Once a protein or lipid is produced, it may not immediately be needed by the cell. Cells can store materials for later use or to secrete in large quantities all at once. If the cell needs to store a product for later, this is typically done in what's called a vacuole. Vacuoles are like vesicles in that they are membrane sacs, but vacuoles are larger and usually stationary. If cells are storing materials for longer periods of time, like triglycerides or proteins, these molecules can be stored in vacuoles and accessed later.

3. What do you think would happen if the ribosomes on the rough endoplasmic reticulum were no longer functional?

4. Why does the nucleus direct the functions of the cell? What's so special about the nucleus that it gets to be the boss?

5. What's the difference between a vesicle and a vacuole?

6. Diagram the process of producing a protein, labeling the organelles involved and adding a short description of what happens at each step.

Maintaining Homeostasis

All organelles are involved in the maintenance of homeostasis within the cell. However, let's focus on a few that are directly involved in maintaining the functioning of organelles within the cell and providing energy so the cell can do all the wide variety of things it needs to do.

Lysosomes, which contain digestive enzymes, manage waste disposal, break down cellular debris and worn-out organelles, and recycle their components for other purposes.

The mitochondria, known as the powerhouse of the cell, generate most of the cell's ATP from the food consumed by the cell. ATP is the chemical energy source that fuels numerous cellular processes, which is critical in maintaining homeostasis. A billion or so years ago, mitochondria were actually bacteria that lived on their own. Their structure and function are a bit unique due to this history, which will be important when we discuss cellular respiration.

All organelles are interconnected, not just spatially but functionally; they form an integrated network where the output of one process in one organelle serves as the input for another. For instance, proteins made in the rough ER are sent to the Golgi apparatus for modification before being shipped to their final destinations. The energy supplied by mitochondria supports these activities, and the entire operation is coordinated by the nucleus, which ensures that gene expression is timely and adjusts the synthesis rates of all cellular materials. This orchestrated working of organelles is fundamental to the cell's ability to maintain internal stability in the face of external changes, the definition of homeostasis.

7. Some diseases can be traced to the malfunctioning of specific organelles. What do you think would be the symptoms of a disease that was traced to the malfunctioning of the mitochondria?

8. If you had to argue one organelle as being the most important to maintaining homeostasis, which would it be and why?

Multiple Choice Review

1. Which statement best describes the role of organelles within a cell?
   a. Organelles limit the cell's ability to metabolize nutrients.
   b. Organelles perform distinct roles necessary to maintain homeostasis and support the cell's functions.
   c. All cells contain the same organelles that perform the same functions.
   d. Organelles are only present in animal cells, not in plant cells.

2. What allows multicellular organisms to perform more complex functions than single-celled organisms?
   a. The absence of DNA in multicellular organisms
   b. The specialization and division of labor among different cell types
   c. The uniformity of cells within multicellular organisms
   d. The larger size of individual cells within multicellular organisms

3. What is a key structural difference between prokaryotic and eukaryotic cells?
   a. Prokaryotic cells are generally larger than eukaryotic cells.
   b. Eukaryotic cells have a defined nucleus, while prokaryotic cells do not.
   c. Prokaryotic cells contain multiple types of organelles, unlike eukaryotic cells.
   d. Eukaryotic cells lack a plasma membrane.

4. Where is the DNA located in a prokarytic cell?
   a. Within the nucleus
   b. Within mitochondria
   c. In a region of the cytoplasm called the nucleoid
   d. Encased within vesicles

5. How are materials such as proteins transported from the rough ER to the Golgi apparatus?
   a. Through direct contact between the rough ER and Golgi
   b. Encapsulated within vesicles that travel via the cytoskeleton
   c. Floating freely through the cytoplasm
   d. Via passive diffusion through the cell

6. Which organelle is responsible for the modification, sorting, and packaging of proteins and lipids?
   a. Mitochondria
   b. Golgi apparatus
   c. Nucleus
   d. Smooth ER

7. How do mitochondria contribute to cellular homeostasis?
   a. By providing structural support to the cell
   b. By producing ATP, the chemical energy needed for various cellular processes
   c. By synthesizing proteins needed by the cell
   d. By recycling worn-out cell organelles

8. What is the primary function of lysosomes within the cell?
   a. To generate ATP
   b. To manage waste disposal by breaking down cellular debris and recycling components
   c. To modify and package proteins
   d. To store nutrients and energy

7

7.2 Cell Detective

Biology textbooks often use very similar cartoons to depict cell structure and shape, as shown in Figure 7.4. However, cells vary widely in their shape, organelle content, and other elements of structure. These differences in structure can be linked to differences in function. For example, muscle cells (usually referred to as "fibers" rather than cells) are elongated and contain large numbers of mitochondria because both of those features improve the cells' ability to contract.

Figure 7.4. Diagram of Prokaryotic and Eukaryotic Cells

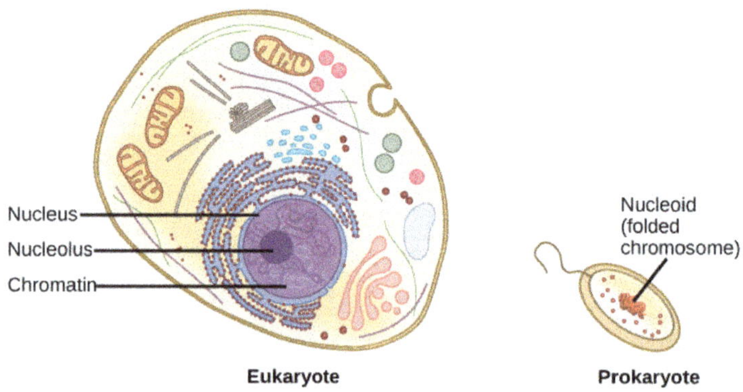

Figure 7.5 is an electron microscope image of a goblet cell. Goblet cells are involved in the secretion of proteins called mucin. In this image, various components of the cell have been circled and labeled. Use this image to answer the questions that follow.

Figure 7.5. A Micrograph of a Goblet Cell

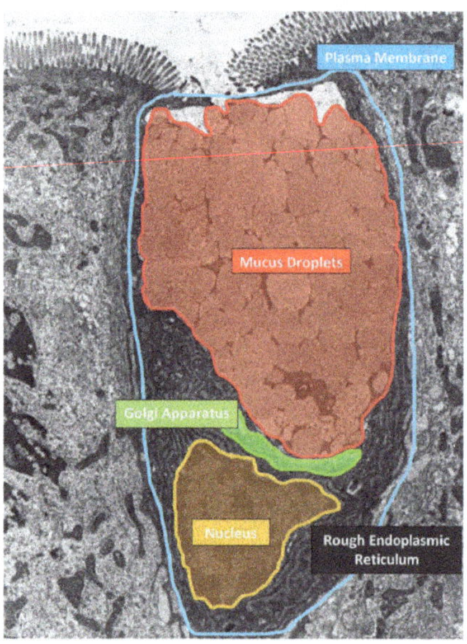

1. Is a goblet cell prokaryotic or eukaryotic? How do you know?

2. Is a goblet cell a plant cell or an animal cell? How do you know?

3. Refer back to the images of prokaryotic and eukaryotic cells. Compare the image of the goblet cell to the cartoon of the cell types that you identified in questions 1 and 2. Describe how the goblet cell differs from the cartoon diagram.

4. Describe how these differences that you identified relate to the function of the goblet cell, which is to produce and release the protein mucin.

7.3 Cell Cycle

All organisms have life cycles. For instance, humans are born, grow and mature through puberty, then reach adulthood. Eventually, our cells and tissues start to break down, and we begin aging. Cells proceed through a life cycle of sorts, as well, which we call the cell cycle. We'll divide up the cell cycle into two parts: interphase and mitosis. Shown below in Figure 7.6 are HeLa cells, at various stages of the cell cycle.

Figure 7.6. HeLa Cells Proceeding Through the Cell Cycle

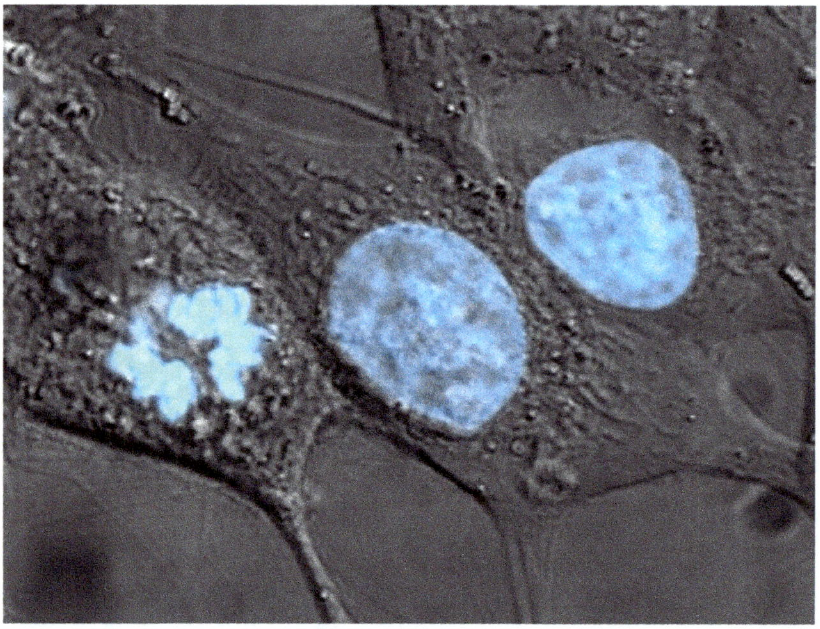

Interphase

Interphase is the part of the cell cycle where most of a cell's time is spent. This is the "life as normal" stage of a cell's life, where it is carrying out its functions and maintaining homeostasis. For instance, if this is a cell in the pancreas whose main job is to produce insulin to maintain blood sugar levels, this pancreatic cell will process the energy necessary to maintain homeostasis and be responsive to signals that tell this cell to produce insulin.

The interphase is divided into three phases: Growth 1 (G_1), Synthesis (S), and Growth 2 (G_2). Interphase literally means "between phase" and implies what happens between mitosis. The cell cycle is diagrammed below in Figure 7.7.

Figure 7.7. A Schematic of the Cell Cycle

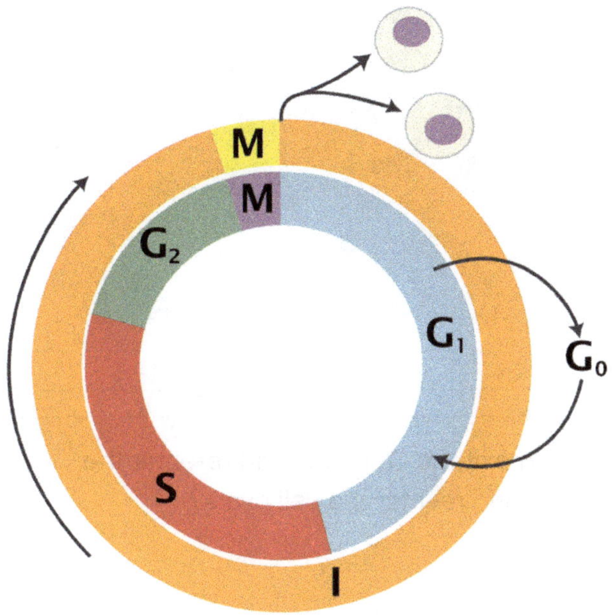

G_1 and G_0

Growth 1 phase, or G_1 phase, is a period of interphase during which the cell undergoes significant growth. This phase marks the beginning of the cell cycle following cell division (mitosis). It is characterized by active production of proteins and other molecules necessary to carry out the cell's functions. These macromolecules are essential for increasing the cell's size and for the accumulation of the necessary components to construct additional organelles needed to efficiently carry out cellular functions. If we're still considering that cell in the pancreas whose role is to create insulin (a protein), this cell would be building up necessary amounts of rough ER so it can efficiently produce insulin during G_1.

During G_1, the cell not only increases in size but also assesses the surrounding environment to ensure it is favorable for replication before proceeding to the S phase, where DNA synthesis occurs. The length of G_1 can vary greatly between cell types and is influenced by both internal regulatory mechanisms and external factors such as nutrient availability and growth factor hormones. If conditions are not optimal and/or the signals for division are not received, the cell can enter a state known as G_0 (pronounced G-not). G_0 is like the cell pressing pause on the cell cycle. In G_0, the cell remains metabolically active but ceases to divide.

This regulation of the cell cycle ensures that the cell only commits to the complete cell cycle and division when conditions are suitable, and division is necessary, thus maintaining proper tissue growth and repair mechanisms. Entering G_0 is reversible for some cells, like liver cells, which may need to divide at a later time. However, some cells, like many neurons, bone, and muscle cells, do not come out of G_0 once they enter it. New types of these cells must be produced from the division of stem cells.

1. Considering the role of the G_1 phase, explain how a cell in the pancreas might adjust its activities during G_1 when there is an increased demand for insulin.

2. Describe a scenario where a liver cell would exit the G_0 phase and re-enter the cell cycle. Why would a cell need to exit G_0 and reenter the cell cycle?

7

S phase

The **S phase, or synthesis phase**, of interphase is part of the cell cycle where DNA replication occurs. The cell will only enter S phase if it receives a signal, like a growth hormone, and if the necessary nutrients are available. During this phase, each chromosome in the cell's nucleus is duplicated to ensure that both resulting daughter cells will receive an identical set of chromosomes during mitosis as shown in Figure 7.8.

The S phase is meticulously regulated to prevent errors in DNA replication that could lead to mutations, as mutations could compromise the integrity of the cell and organism. Completing this phase successfully is essential for the cell to progress to the next stages of the cell cycle, G_2 and mitosis, where further preparation for cell division occurs and then the physical separation of the cell's components into two daughter cells.

Figure 7.8. Duplication of Chromosomes

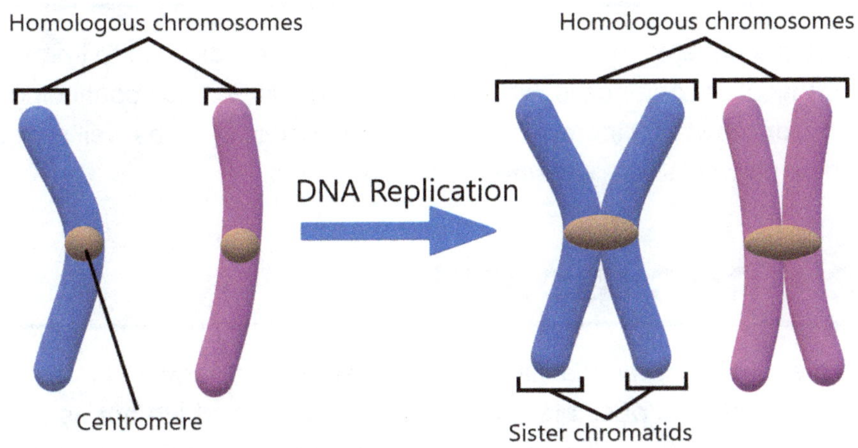

G_2

The **G_2 phase** of interphase occurs right after DNA synthesis (S phase) and before the cell enters mitosis. During this phase, the cell grows rapidly in order to prepare for cell division. This preparation includes the production of proteins and organelles so the resulting two daughter cells will have everything they need to successfully get started. This phase also involves substantial checks and repairs to ensure that the DNA has been replicated correctly without any damage or errors. The cell also synthesizes cytoskeleton microtubules necessary for forming the mitotic spindle, which will be used to separate sister chromatids during mitosis.

3. Why does the cell need to replicate its DNA?

4. Would a normally functioning cell enter G_2 phase if S phase had not been successfully completed? Explain why you think this.

Mitosis

The details of cell division are covered in a later section. For now, we just need to know that mitosis produces two identical daughter cells, which each start out in G_1, thus continuing the life cycle. Mitosis is used for growth by increasing the total number of cells, as well as repair of damaged cells, like how new cells are generated to heal a wound.

Cell Cycle Checkpoints

Cell cycle checkpoints are regulatory mechanisms that ensure the accuracy and integrity of cell division. These checkpoints are possible due to sets of proteins which act as surveillance systems that assess whether the cell is ready to proceed through the various stages of the cell cycle—such as DNA replication, chromosome segregation, and division completion. Positioned strategically at key points in the cell cycle, as shown in Figure 7.9, particularly before entering the S phase (DNA synthesis), before mitosis (M phase), and during metaphase, these checkpoints help detect and repair DNA damage, ensure all chromosomes are properly replicated and attached to the spindle apparatus, and confirm that all cellular components are suitably prepared for division. If errors or damages are detected, the checkpoint mechanisms can pause the cell cycle, allowing the cell time to repair the damage or, in cases where damage cannot be repaired, to initiate programmed cell death (apoptosis) to prevent the passage of faulty genetic information to daughter cells. This system is essential for preventing diseases such as cancer, where unchecked cellular division can lead to tumor growth and the spread of malignant cells throughout the body.

Figure 7.9. Cell Cycle Checkpoints

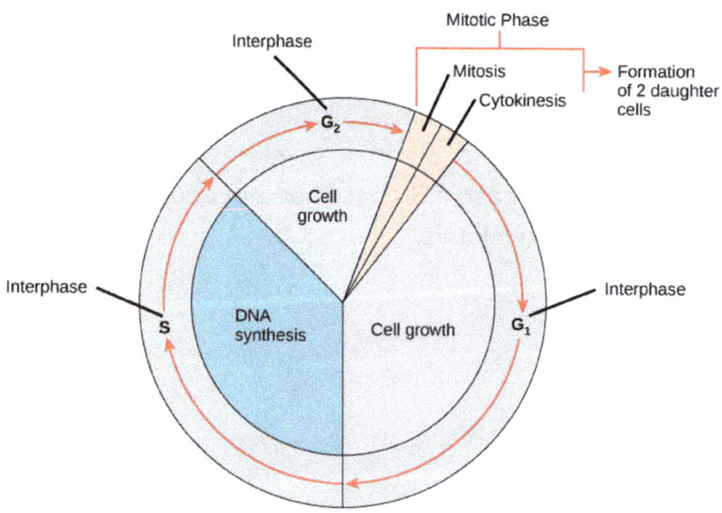

5. What is the purpose of cell cycle checkpoints?

6. What could be the result if cell cycle checkpoints are not abided by?

Multiple Choice Review

1. How does the cell cycle contribute to tissue growth and repair?
   a. By enabling cells to divide and replace damaged or old cells
   b. By allowing cells to reduce their metabolic activities
   c. By causing cells to increase their size only
   d. By moving cells to different tissues

2. What is the significance of the G0 phase in the cell cycle?
   a. It is a phase where the cell divides rapidly.
   b. It is a resting phase where the cell remains metabolically active but does not divide.
   c. It is when the cell is actively replicating its DNA.
   d. It is when the cell prepares for immediate death.

3. What is the role of checkpoints in the cell cycle?
   a. To ensure the cell divides as quickly as possible
   b. To monitor and verify whether the processes at each phase have been accurately completed before progressing to the next phase
   c. To stop all cell functions except division
   d. To initiate the process of apoptosis

4. During which phase does DNA replication occur?
   a. G_1 phase
   b. S phase
   c. G_2 phase
   d. M phase

8.1 Mitosis & Meiosis

Mitosis

Mitosis is a fundamental process for cell division that ensures the equal distribution of genetic material into two daughter cells. It is a meticulously orchestrated sequence of steps, each marked by distinct structural changes in the cell and regulatory mechanisms to coordinate the process of dividing DNA equally into the two daughter cells. The primary function of mitosis is to enable the growth and repair of damaged cells in multicellular organisms by producing new cells that are genetically identical to the parent cell. Mitosis divides the sister chromatids formed during the S phase of interphase into two daughter cells, as seen in Figure 10.1.

Figure 10.1. Diagram of Homologous Chromosomes Before and After DNA

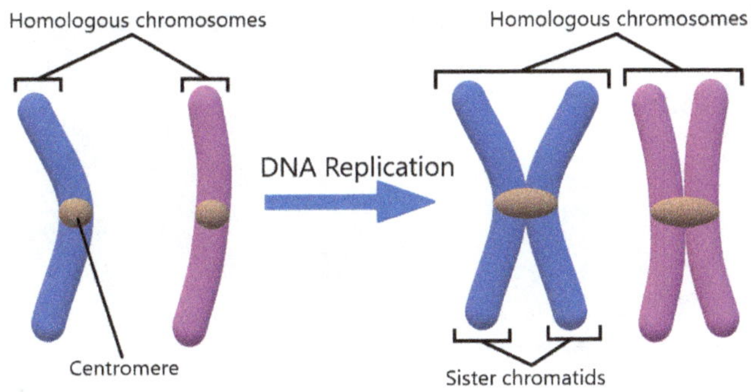

1. Is DNA replicated before or after mitosis begins? Explain when in the cell cycle DNA replication occurs.

Phases of Mitosis

The mitotic process is divided into several key phases: prophase, metaphase, anaphase, and telophase, each characterized by specific events and structures. In **prophase**, chromosomes condense and become visible under a light microscope, and the **mitotic spindle** begins to form as centrosomes move to opposite poles of the cell.

Metaphase follows, where chromosomes align at the cell's equatorial plane, guided by the spindle fibers attached to their centromeres, ensuring each new cell will receive an identical set of chromosomes. **Anaphase** is marked by the separation of sister chromatids, pulled apart by the spindle fibers towards opposite poles, effectively halving the genetic material. Finally, in **telophase**, nuclear membranes reform around the separated chromatids, now considered individual chromosomes, while the cell begins the process of **cytokinesis**, splitting the cytoplasm to form two genetically identical daughter cells. This sequential progression through the mitotic phases ensures the fidelity of cell division, preserving the integrity of the cell's genome across generations, as seen in Figure 10.2.

Figure 10.2. Stages of the Cell Cycle

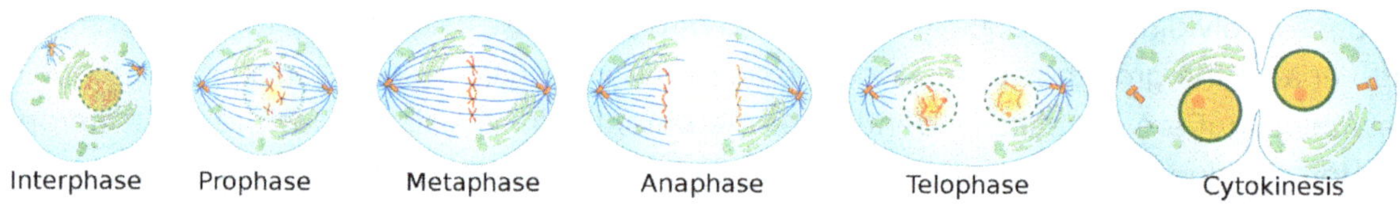

2. Can you explain why the alignment of chromosomes at the cell's equatorial plane during metaphase is crucial for ensuring that each new cell receives an identical set of chromosomes?

3. How does the formation of the mitotic spindle in prophase contribute to the overall process of mitosis and the accurate distribution of genetic material?

4. Why is cytokinesis, following telophase, important for the completion of mitosis, and what would happen if cytokinesis did not occur after the nuclear division?

8

Mitotic Checkpoints

Mitotic checkpoints serve as surveillance mechanisms that ensure the accurate division and integrity of the cell's genetic material during mitosis. These checkpoints, strategically placed at various stages of the cell cycle, assess whether the cell is ready to proceed to the next phase, particularly focusing on chromosome alignment, attachment to spindle fibers, and complete segregation of chromosomes. If these checkpoints are not met, they trigger a halt in the cell cycle, allowing time for repair or, in cases of irreparable damage, leading the cell toward programmed cell death (**apoptosis**) to prevent the propagation of errors.

However, when these checkpoints fail or are bypassed due to mutations in the genes responsible for their regulation, uncontrolled cell division ensues, leading to the accumulation of genetic abnormalities and the potential formation of cancer. This loss of regulatory control allows cancerous cells to multiply without restraint, forming tumors and invading surrounding tissues, a hallmark of cancer's progression and malignancy. The integrity of these checkpoints is thus paramount in preventing the onset of cancer, highlighting the delicate balance the cell must maintain to ensure its proliferation does not lead to pathological consequences, as seen in Figure 10.3.

Figure 10.3. Overview of the Cell Cycle Checkpoints and Phases

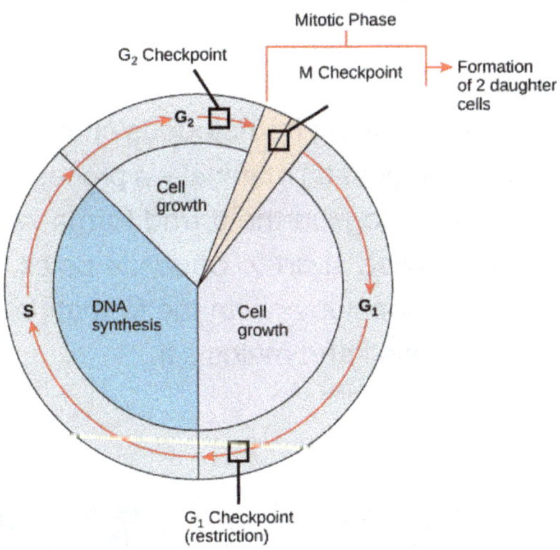

5. Several drugs specifically target and impact the formation of spindle fibers during mitosis, primarily affecting the microtubules that make up the spindle apparatus. These drugs are often used in cancer treatment. Why would they be useful in treating cancer?

Meiosis

Meiosis is an essential cellular process that divides a diploid cell into four haploid cells, a quintessential step in sexual reproduction that contributes to genetic diversity. This complex form of cell division comprises two sequential rounds of division—meiosis I and meiosis II—each with its own series of phases, similar to mitosis, but with a distinctive purpose. The primary function of meiosis is to halve the number of chromosomes, ensuring that each gamete, sperm, or egg contains just one set of chromosomes.

This reduction is fundamental to sexual reproduction, allowing for the restoration of the diploid state upon fertilization. During meiosis, homologous chromosomes undergo recombination, shuffling genetic information and creating new combinations of alleles, which increases genetic variation within a species. Such diversity is vital for evolution and the adaptability of organisms to changing environments.

The steps of meiosis are similar to mitosis, with a couple of major differences. The first major difference is we have two rounds of division in meiosis. In the first round, the homologous chromosomes separate. In the second round of divisions, the sister chromatids separate. Here's a hint: if the phases are numbered, that indicates we're talking about a step of meiosis. We don't number the phases of mitosis.

8

Meiosis I, as seen in Figure 10.4, initiates with **prophase I**, where homologous chromosomes pair up, which forms a **tetrad**. Once paired, they exchange genetic material—a process known as **crossing over**. Metaphase I follows, with paired homologs in the tetrad formation aligning at the cell's equator. In anaphase I, these pairs are pulled apart to opposite poles, reducing the chromosome number by half. Finally, telophase I concludes with the formation of two new nuclei. There is no replication of DNA between meiosis I and meiosis II.

Figure 10.4. Stages of Meiosis I

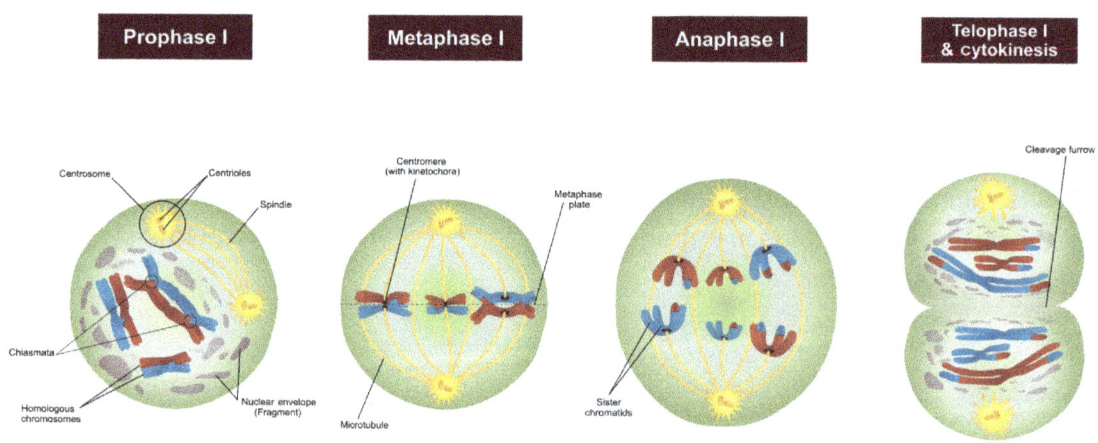

The chromosomes condense, and the nuclear envelope breaks down. Crossing-over occurs.

Pairs of homologous chromosomes move to the equator of the cell.

Homologous chromosomes move to the opposite poles of the cell.

Chromosomes gather at the poles of the cells. The cytoplasm divides.

Meiosis II, as seen in Figure 10.5, is similar to mitosis: in prophase II, the chromosomes condense again; during metaphase II, chromosomes align singly; in anaphase II, sister chromatids segregate to opposite poles; and telophase II leads to the formation of four genetically distinct haploid cells. Each phase is critical, with mechanisms ensuring genetic diversity and the correct number of chromosomes in gametes, thereby playing an instrumental role in the survival and evolution of species.

Figure 10.5. Stages of Meiosis II

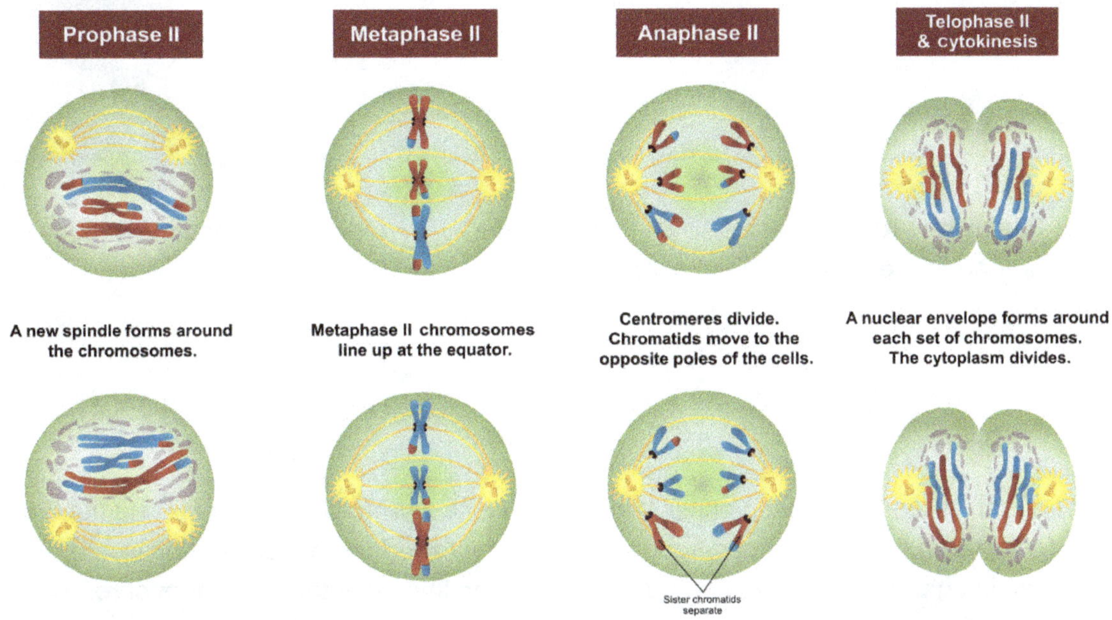

6. How does meiosis differ from mitosis in terms of the number of divisions and the outcome in terms of cell number and chromosome content?

7. What is the significance of homologous chromosomes undergoing recombination during meiosis, and how does this process contribute to genetic diversity?

8. Explain the role of crossing over in meiosis I and its impact on genetic variation among the resulting haploid cells.

Meiosis is an essential cellular process that underpins sexual reproduction, enabling the transfer of genetic information across generations. This specialized form of cell division is responsible for producing gametes—sperm and eggs—each containing half the number of chromosomes found in a normal body cell. Such a reduction is crucial, as it allows for the restoration of the full chromosome complement upon fertilization, preserving the species' characteristic chromosome number.

During meiosis, homologous chromosomes undergo a unique exchange of genetic material through a process called crossing over, which fosters genetic diversity. This recombination is the physical manifestation of the genetic shuffling that contributes to the variation seen within a species. Meiosis not only ensures the continuity of genetic traits from parent to offspring but also introduces novel genetic combinations into the population, a driving force behind the evolution of species and the endless genetic variation of individuals.

Gametes, specifically sperm and ova, are essential for inheritance and play a direct role in the patterns of reproduction. Produced through meiosis, these haploid cells carry half the chromosome number of normal cells, ensuring genetic variety when they combine during fertilization. This recombination results in a zygote with a complete set of chromosomes, where the genotype—the genetic makeup—is established. The genotype determines the potential for various traits, known as the phenotype, which will develop under specific environmental influences, as seen in Figure 10.6.

Figure 10.6. Human Life Cycle

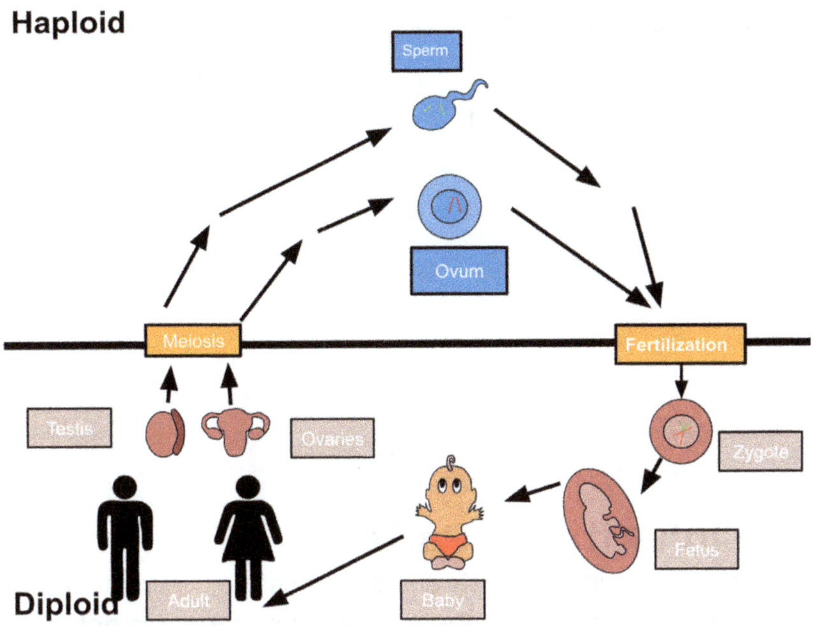

Gametes play a pivotal role in the genetic transfer from one generation to the next, adhering to the principles of Mendelian genetics. **Segregation** is the process where pairs of gene variants (alleles) are separated into different gametes during meiosis, ensuring that each gamete carries only one allele for each gene (the result of Meiosis I). This explains why offspring have a 50% chance of inheriting either allele from a single parent. To add to the diversity, gametes help create **independent assortment,** which describes how different genes independently separate from one another when reproductive cells develop.

This means the distribution of one pair of alleles into gametes does not influence the distribution of another pair as long as they're not on the same chromosome. The practical outcome of these principles is a diverse combination of alleles in the offspring, a process that is central to the understanding of inheritance patterns. The ability to predict the distribution of traits relies on these genetic mechanisms, which serve as the bedrock for classical genetics and the study of heredity.

9. In the context of meiosis and gamete production, explain the significance of reducing the chromosome number by half in gametes and how this affects genetic inheritance and the restoration of the chromosome number upon fertilization.

10. Describe the roles of segregation and independent assortment in meiosis and how they contribute to the diversity of genetic combinations in offspring, including the impact on the distribution of alleles.

8

Dominant-Recessive Inheritance Patterns

Due to the separation of sister chromatids during meiosis and the process of fertilization, simple dominant-recessive patterns can emerge, though these types of patterns are not often seen in human traits. This pattern hinges on the relationship between two versions of a gene—**alleles**—one dominant and one recessive. The **dominant allele**, if present, expresses its trait even in the company of a contrasting recessive allele. The recessive trait only emerges in the phenotype—the observable characteristics of an organism—when it is paired with another **recessive allele**, devoid of the dominant counterpart's influence.

When examining an individual's genotype, or their combination of alleles, the presence of a dominant allele (represented typically by a capital letter, for example, 'A') in either homozygous (AA, homozygous dominant) or heterozygous (Aforms results in the dominant phenotype. The recessive phenotype only appears when both alleles are recessive (aa or homozygous recessive). This foundational concept of inheritance explains the predictable patterns seen in various traits across generations. It's this dominant-recessive inheritance that guides genetic predictions, from predicting the likelihood of a child having a family trait to understanding the transmission of certain genetic conditions.

A Punnett square, as seen in Figure 10.7, is a simple graphical method used to predict the outcomes of a genetic cross. To create one, draw a grid with four squares. On the top, write the alleles (gene variants) of one parent, and on the side, write the alleles of the other parent. If a parent is heterozygous (having two different alleles for a trait, like Aa), you'll write one allele above one column and the other allele above the adjacent column. Do the same on the side for the second parent. Then, fill in each square by combining the alleles from the top and the side, showing all possible combinations for the offspring. This visual tool helps in understanding the genetic makeup of offspring, predicting trait distributions, and calculating the probability of inheriting specific traits. It's particularly useful for simple genetic crosses involving one or two traits and follows Mendelian inheritance patterns.

Figure 10.7. Punnett Square

	A	*a*
A	AA	Aa
a	Aa	aa

For example, if we consider a gene where the dominant allele codes for brown eyes (and the recessive allele codes for blue eyes (b), a person with at least one dominant allele will have brown eyes. It's only when an individual inherits two recessive alleles (bb), one from each parent, that they will have blue eyes. This is not actually how the inheritance of eye color works in humans, but it's simply an example.

11. Where does each allele in an individual's genotype come from? (hint: think about meiosis and reproduction)

12. How does the presence of dominant and recessive alleles in an individual's genotype determine their phenotype?

13. If curly hair (is dominant to straight hair (c), and one parent is homozygous dominant for curly hair (CC) while the other parent is heterozygous (Cc), use a Punnett square to calculate the chances of their offspring having straight hair. Describe the steps you take to fill out the Punnett square and the likelihood of each possible genotype.

Multiple Choice Review

1. What is the primary purpose of mitosis in multicellular organisms?
   a. To decrease the chromosome number in gametes
   b. To produce genetically diverse offspring
   c. To replicate cells for growth, repair, and maintenance of the organism
   d. To exchange genetic material between homologous chromosomes

2. During which phase of mitosis do the sister chromatids separate and move to opposite poles of the cell?
   a. Anaphase
   b. Prophase
   c. Metaphase
   d. Telophase

3. What is the significance of the spindle fibers during mitosis?
   a. They replicate the cell's DNA.
   b. They align chromosomes at the cell's equator during metaphase.
   c. They produce the energy needed for cell division.
   d. They decrease the cell's volume.

4. A scientist observes a cell with its chromosomes aligned at the center of the cell. Which phase of mitosis is the cell most likely in, and what will happen next?
   a. Prophase: the nuclear envelope will dissolve.
   b. Anaphase: the chromosomes will de-condense and form two new nuclei.
   c. Metaphase: the sister chromatids will be pulled to opposite poles.
   d. Telophase: the cell will begin cytokinesis.

5. If a cell fails to properly attach all chromosomes to the spindle fibers during mitosis, which checkpoint is most likely to halt the cell cycle until the issue is resolved?
   a. G_2 checkpoint
   b. G_1 checkpoint
   c. Metaphase checkpoint
   d. Anaphase checkpoint

6. In an experiment, a cell is treated with a chemical that inhibits the formation of the spindle fibers. What stage of mitosis would be directly affected by this treatment, and what would be the likely outcome?
   a. Prophase: chromosomes would not condense.
   b. Anaphase: sister chromatids would not separate.
   c. Telophase: nuclear envelopes would not form around the separated chromosomes.
   d. Metaphase: chromosomes would not align at the cell's equator.

8

7. How does meiosis contribute to genetic diversity in sexually reproducing organisms?
   a. By replicating the chromosomes before cell division
   b. Through the processes of crossing over and independent assortment
   c. By producing identical daughter cells
   d. By allowing sister chromatids to remain together

8. During which stage of meiosis do homologous chromosomes pair up and form tetrads?
   a. Metaphase I
   b. Anaphase I
   c. Telophase I
   d. Prophase I

9. What is the outcome of meiosis in terms of the number and type of cells produced?
   a. Two genetically identical diploid cells
   b. Two genetically diverse diploid cells
   c. Four genetically identical haploid cells
   d. Four genetically diverse haploid cells

10. Given a scenario where a plant species has a diploid chromosome number of 6 (2n=6), how many chromosomes will each gamete contain after meiosis?
   a. 3 chromosomes
   b. 6 chromosomes
   c. 12 chromosomes
   d. 24 chromosomes

11. In a plant species, tall (T) is dominant over short (t). A heterozygous tall plant (Tt) is crossed with a short plant (tt). What is the probability of producing a short plant from this cross?
   a. 0%
   b. 25%
   c. 50%
   d. 75%

12. If a child exhibits freckles (F), a dominant trait, and their mother does not have freckles (ff), which of the following could be the possible genotype(s) of the child's father?
   a. Homozygous dominant (FF)
   b. Homozygous recessive (ff)
   c. Heterozygous (Ff)
   d. Both A and C are possible

9.1 Nucleic Acids

Nucleic acids are the information molecules of cells. While they have some secondary functions, nucleic acid polymers mainly function to store and transmit genetic information. A large portion of the information encoded by nucleic acids is the instructions for how to build proteins. The function of nucleic acids and the process of how they are used to transmit information and direct the building of proteins will be covered later. Here, we're mainly focused on the structure of the molecule and its general function in cells.

Nucleotides

Nucleotides are the monomers of nucleic acids, including DNA (deoxyribonucleic acid) and RNA (ribonucleic acid), which are essential macromolecules for all forms of life. The standard nucleotide structure is shown in Figure 9.1. Each nucleotide consists of three components:

- **Nitrogenous Base:** These are organic molecules and include purines (adenine [A] and guanine [G]) and pyrimidines (cytosine [C], thymine [T] in DNA, and uracil [U] in RNA). These bases are critical for the encoding of genetic information, as their sequence forms the genetic code.
- **Five-Carbon Sugar:** In DNA, this sugar is deoxyribose, which lacks one oxygen atom present in the ribose sugar of RNA. This structural difference is crucial for the stability of DNA and the more transient nature of RNA.
- **Phosphate Group:** This component links the nucleotides together through phosphodiester bonds, forming the backbone of the nucleic acid strand. The phosphate is bonded to the 5' carbon of one sugar and the 3' carbon of the next sugar, creating a directionality to the nucleic acid strand from 5' to 3'.

Figure 9.1. Components of Nucleotides

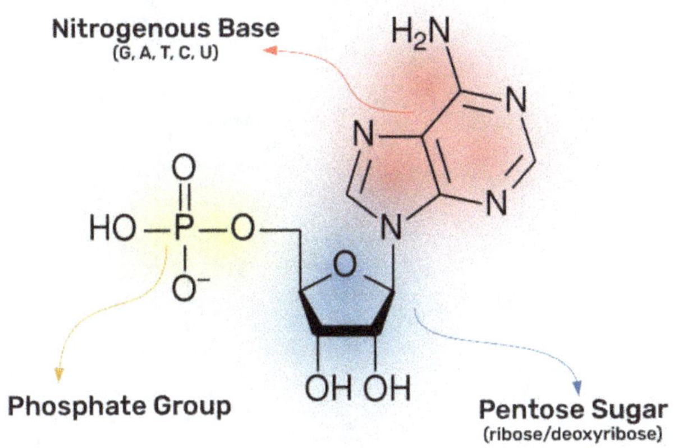

1. Explain the role of nitrogenous bases in nucleotides and how they contribute to genetic information encoding.

2. What reaction would connect nucleotides together into a polynucleotide? What reaction would break down a polynucleotide into its monomers?

RNA vs DNA

In DNA, the sequence of nitrogenous bases along the nucleotide polymer encodes the genetic information necessary for the synthesis of proteins, which are crucial for the structure, function, and regulation of the body's tissues and organs. The specific pairing between the bases (adenine with thymine, cytosine with guanine in DNA, adenine with uracil in RNA) facilitates the accurate replication and transcription of genetic information. The double stranded nature of DNA is shown in Figure 9.2.

Figure 9.2. Structure of DNA

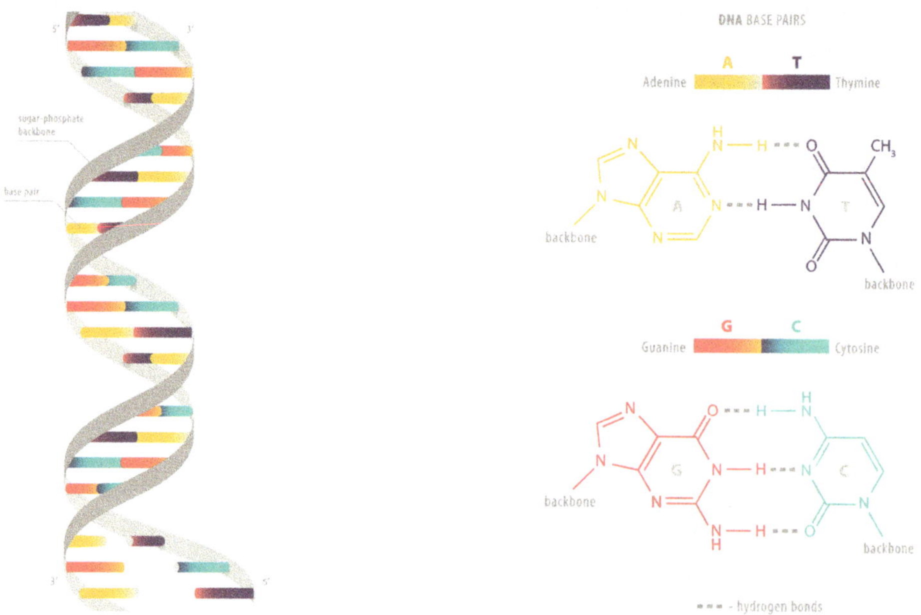

The differences summarized in Table 9.1 allow DNA and RNA to fulfill their specialized roles in the cell. DNA acts as a durable storage medium for genetic information, and RNA serves multiple functions in the expression and regulation of these genetic instructions.

Table 9.1.

	DNA	RNA
Structure	DNA is a double-stranded molecule that forms a double helix, providing a stable structure for the long-term storage of genetic information.	RNA is typically single-stranded, which allows it to fold into complex shapes and perform various functions in the cell, such as catalysis, signaling, and as a template for protein synthesis.
Sugar	Contains deoxyribose, a sugar with one fewer hydroxyl group (-OH) compared to ribose. The absence of an oxygen atom at the 2′ position of the sugar ring makes DNA more chemically stable.	Contains ribose, which has a hydroxyl group at the 2′ position of the sugar ring, making RNA more reactive and less stable than DNA.
Bases	Uses adenine (A), guanine (G), cytosine (C), and thymine (T) as its four nitrogenous bases. A and T are complementary, and C and G are complementary.	Uses adenine (A), guanine (G), cytosine (C), and uracil (U) instead of thymine. A and U are complementary, and C and G are complementary.
Location	In eukaryotic cells, DNA is located primarily in the cell nucleus.	Found throughout the cell, including the nucleus, cytoplasm, and within the ribosomes of the endoplasmic reticulum.

9

The Genome

The organization of the eukaryotic genome is a hierarchical structure designed to manage and compactly store vast amounts of genetic information. All the instructions to build any part of a cell, of all cell types, is encoded in the DNA found in the nucleus of each eukaryotic cell. This is a large amount of information that needs to be stored in a very small space. The organization of the genome ensures that information can be efficiently accessed, expressed, or silenced as needed by the cell. Figure 9.3 shows a focused overview of how the genome is organized in Eukaryotic cells.

Figure 9.3. Organization of the Genome

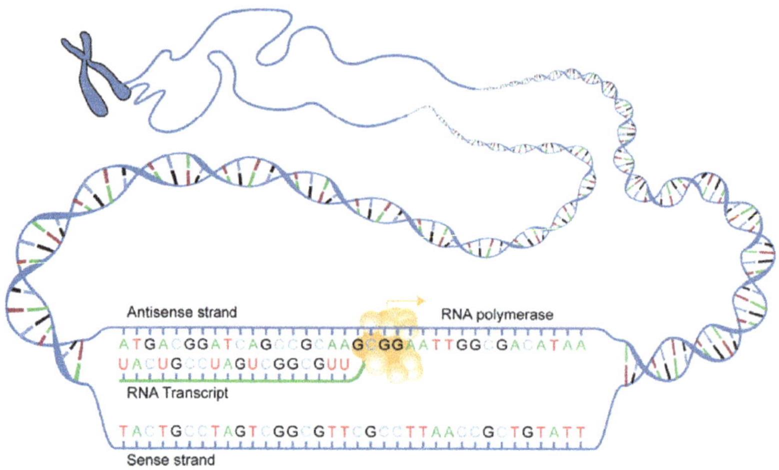

DNA and Histones

At the most basic level, DNA is wrapped around histone proteins to form chromatin. At this level, DNA looks like beads on a string, with the DNA double helix winding around histone proteins. The histone proteins function to compact the DNA into a smaller, more manageable shape while maintaining access to the DNA molecule. Chromatin is the combination of DNA and protein (histones and others) found in the nucleus. Chromatin is the material that makes up chromosomes.

9

Chromosomes

Chromatin further folds and condenses to form chromosomes, which are long, linear DNA molecules associated with proteins. Humans have 23 pairs of chromosomes, each containing a single piece of DNA that holds thousands of genes. The chromosomes are where DNA is organized and stored in the nucleus.

Genes

Within chromosomes, genes are segments of DNA that code for proteins or RNA molecules. How genes function to produce protein, and the different components of a gene, will be covered later.

Base pairing

The base pairing rule of DNA explains how nucleotides form pairs across the two strands of the DNA helix. In DNA, adenine (A) pairs exclusively with thymine (T) through two hydrogen bonds, and guanine (G) pairs with cytosine (C) through three hydrogen bonds. This specific pairing ensures the DNA structure's stability and fidelity during replication and transcription processes.

3. Describe the role of histones in the organization of DNA within the nucleus of eukaryotic cells.

4. If humans have 23 pairs of chromosomes, how many total chromosomes are in a typical cell? Do all cells have the same amount of chromosomes? Explain your thinking.

5. If one strand of DNA has the sequence AATGCCGTAT, what would the sequence of the complementary strand be?

DNA Replication

During reproduction, a cell will need to replicate its DNA, or make an exact copy of its DNA, so that each new cell that is produced gets a full copy of the entire genome. DNA replication is the process of creating new DNA from existing DNA. This process is called "semiconservative" because each new DNA molecule consists of one original strand and one newly synthesized strand on each half of the double helix, as shown in Figure 9.4.

Figure 9.4. Semi-Conservative Model of DNA Replication

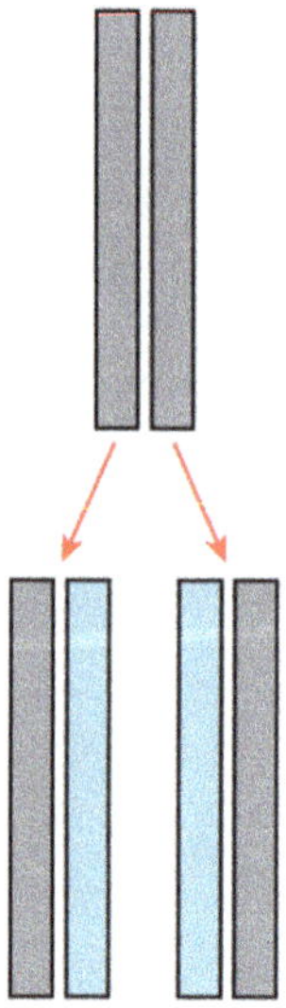

Unwinding

First, the DNA double helix unwinds as seen in Figure 9.5. An enzyme called helicase breaks the Hydrogen bonds between the nitrogenous bases of the two strands of DNA, separating them.

Figure 9.5. Replication of DNA

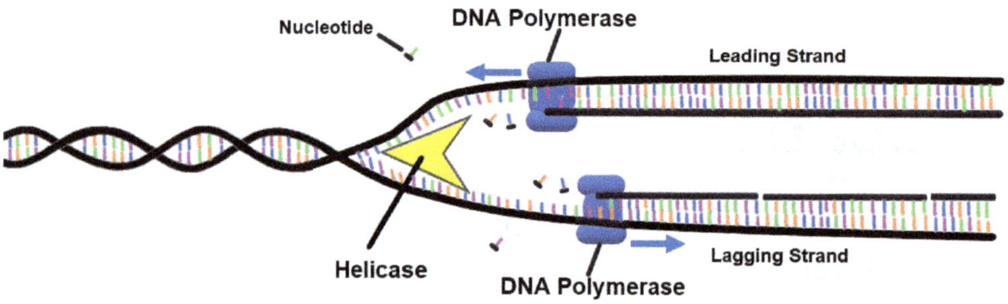

Building New Strands

The synthesis of new DNA is carried out by the enzyme called **DNA polymerase**. The name of the enzyme tells us what it does, so let's look at what its name means. A polymer is many monomers connected together. The -ase ending indicates it's an enzyme. The "DNA" in front of "polymerase" tells us we're working specifically with DNA. So, DNA polymerase is an enzyme that builds polymers of DNA. DNA polymerase can "read" the original strand of DNA, and add a complementary base to the newly synthesized strand at that location, as diagrammed in Figure 9.5. The nucleotides are connected together by DNA polymerase catalyzing the bond between the phosphate group of one nucleotide and the sugar of the next nucleotide, making a covalent bond that forms the sugar-phosphate backbone of DNA.

For one strand (the leading strand), this building process goes smoothly and continuously because it's in the right direction for the DNA polymerase to work. For the other strand (the lagging strand), the process is a bit choppy. DNA polymerase can only work in one direction, so it has to keep jumping back as the unwinding reveals more of the template strand, creating short pieces that will be connected later.

At the end, you have two DNA molecules. Each one has one of the original strands and one new strand. This method ensures that each new cell gets a perfect copy of the DNA, ready for action. Having one half of the DNA molecule act as a template for the new DNA ensures that the sequence of bases is maintained during replication, which minimizes the risk of errors (called mutations).

6. Explain the role of DNA polymerase in the process of DNA replication.

7. What is the benefit of having one strand of DNA act as a template for the newly synthesized DNA?

Multiple Choice Review

1. How does the structure of RNA differ from DNA in terms of sugar components?
   a. RNA contains deoxyribose, lacking one oxygen atom.
   b. DNA and RNA both contain ribose.
   c. RNA contains ribose, which has one more oxygen atom than DNA's deoxyribose.
   d. RNA uses a completely different sugar than DNA.

2. What is the primary function of nitrogenous bases in nucleotides?
   a. To provide energy to cells
   b. To encode genetic information
   c. To connect nucleotides together
   d. To stabilize the DNA structure

3. Which statement best describes a key function of RNA?
   a. RNA is primarily responsible for the long-term storage of genetic information.
   b. RNA primarily serves to replicate and repair genetic material.
   c. RNA acts as a messenger carrying instructions from DNA for controlling the synthesis of proteins.
   d. RNA enhances the structural stability of the cell nucleus.

4. What is the primary function of DNA in cells?
   a. To assist in protein synthesis
   b. To store and transmit genetic information
   c. To act as a catalyst in biochemical reactions
   d. To transport amino acids to ribosomes

5. Which nucleotide pairs with guanine according to the base pairing rules of DNA?
   a. Adenine
   b. Thymine
   c. Cytosine
   d. Uracil

9.2 Gene Expression

What does it mean to "express" something? When we express emotions, we turn our internal emotions into a physical manifestation—crying, smiling, laughing—something that others can see and interact with. When we talk about the expression of genes, it's somewhat similar. Our cells take something internally coded—their DNA—and convert it to a physical product, a protein that can interact with other molecules in the cellular world, as seen in Figure 9.6.

This process of gene expression has two main steps that we'll focus on. The first being transcription, the second being translation. By the end of this module, you should be able to transcribe and translate a gene into its protein product, and identify the roles various molecules play in the process.

Figure 9.6. The Central Dogma of Information Flow in Biology

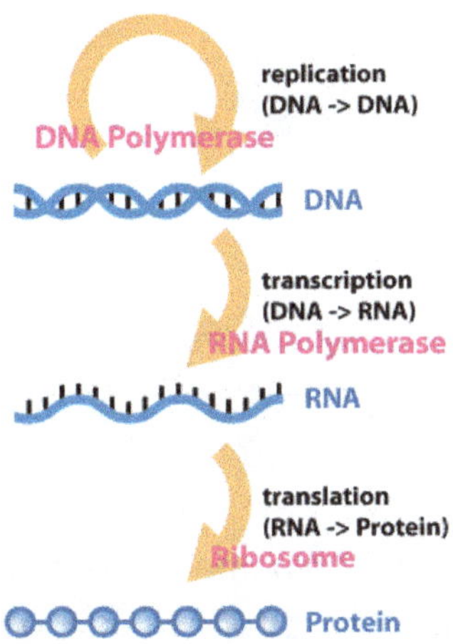

1. What is the purpose of gene expression?

2. What is the end product of gene expression?

Transcription

What is a transcript? Where else have you heard this word before? Fans of true-crime may have heard this term to describe a written version of dialogue that occurred in a courtroom or a court transcript. The court transcriber is a person who documents everything that everyone in the courtroom says during a trial so there is a written copy of the proceedings that can be referenced later. If you've ever taken notes during a lecture, you've made a transcript of the lecture, though in your own words and probably not an exact copy of *everything* the instructor said. The transcript that we're talking about here is most similar to your notes, a truncated version of the original discussion. When talking about gene expression, the process of **transcription** creates an mRNA (messenger RNA) copy of a gene. The entire DNA isn't copied letter for letter, but just the section of DNA that's necessary to build a specific protein, as seen in Figure 9.7.

Figure 9.7. DNA Transcription

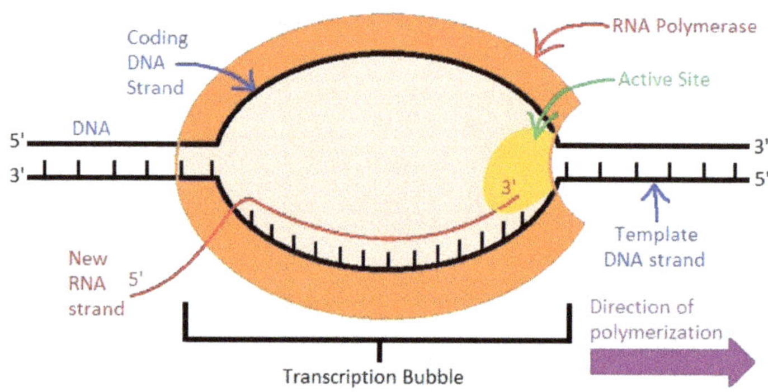

Transcription is carried out by the enzyme called **RNA polymerase**. The name of the enzyme tells us what it does—so let's break it down. A polymer is many monomers connected together. The -ase ending indicates it's an enzyme. The "RNA" in front of "polymerase" tells us we're working specifically with RNA. So, RNA polymerase is an enzyme that builds polymers of RNA.

The process of transcription is much more complex than we'll dive into here, so if you finish this module and feel like this is too simple to be how it really works, you're right. For our purposes, we're going to hit the highlights and leave the rest of the complexities for a future genetics class (if you choose to take any).

Transcription of a gene begins at a location on the DNA called the **promoter**. This is a regulatory region of DNA that helps RNA polymerase figure out where to begin its job of building an mRNA copy of the gene. Once RNA polymerase has bonded to the promoter, it can open up the DNA strand by breaking the hydrogen bonds between about 12 nucleotides at a time. Once those bases are separated, RNA polymerase "reads" the template strand of DNA and builds a single strand of RNA that is complementary to the DNA.

RNA polymerase will continue to move along the gene, opening up a few bases at a time and building the complementary RNA strand, until it comes to a sequence of DNA called the terminator sequence. At this point, RNA polymerase dissociates from the DNA, and the RNA transcript floats away. Transcription is now complete, but we can't call the molecule mRNA yet until it has been processed in the next step.

3. What enzyme is responsible for making an mRNA transcript of a gene?

4. How does the enzyme from the question above know where to start and stop transcription?

5. Where in the cell does transcription take place?

6. What would happen if a cell was unable to transcribe a gene for a critical enzyme, like the enzymes in lysosomes?

RNA Processing

In eukaryotic organisms, the RNA copy of the gene that was made during transcription needs to be edited before it can be used as instructions to build a protein. This step is kind of like writing out a long text message in your notes and then editing it before actually sending it. You might take some pieces out of the draft and change some wording before you actually send the message. In RNA processing, parts of the RNA that will not be used in the final building of

9

the protein are removed, and the remaining pieces are spliced together. The pieces that are removed are called **introns**, and the pieces that remain and are spliced together are called **exons**.

On the 3' end of the RNA, several adenine nucleotides are added, this, several adenine nucleotides are added; this is called a poly-A tail and helps with getting the mRNA out of the nucleus and the general stability of the mRNA once it's in the cytoplasm. On the other end of the RNA, one guanine nucleotide (G) is added to the 5' end and is referred to as the **5' G cap**. The G cap is essential for the initiation of translation and ribosome binding. All of this processing of the RNA into mRNA is carried out by a complex of enzymes collectively called the spliceosome. Once the editing is complete, the mRNA is shuttled out of the nucleus into the cytoplasm, and the next step, translation, will begin, as seen in Figure 9.8.

Figure 9.8. RNA Processing

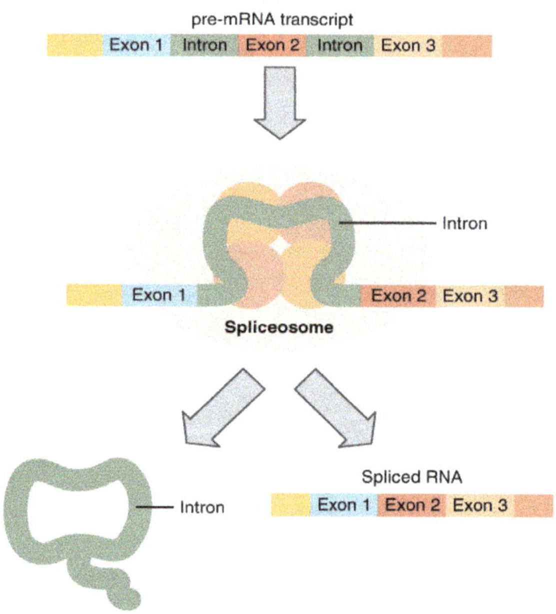

9

7. What is the difference between introns and exons?

8. What other edits are made to the mRNA before translation? What is the purpose of these edits?

Translation

Transcription produces an mRNA copy of a gene. It might be helpful to think of this mRNA copy as a photocopy of an important historical document, like the Bill of Rights. It wouldn't be wise to pass around the original copy of the Bill of Rights (circa 1789), as this would damage an important historical document and potentially destroy the original. Instead, we have copies and reproductions made, which allows us to read the Bill of Rights, and apply it to our lives, but keeps the original version from accumulating damage. In this analogy, the DNA is the original version, and mRNA is the reproduction of the original.

We left off with mRNA having completed its processing, and it was being shuttled out of the tightly controlled environment into the cytoplasm. It's here in the cytoplasm that **translation** of the message the mRNA is carrying in its sequences of bases into a polypeptide chain will occur. The organelle that we learned about in week 6, which is responsible for building proteins, is the ribosome. But, the ribosome will need additional help in order to build the polypeptide, so let's meet the entire cast before we discuss the process of translation, as seen in Figure 9.9.

Figure 9.9. Protein Synthesis Overview

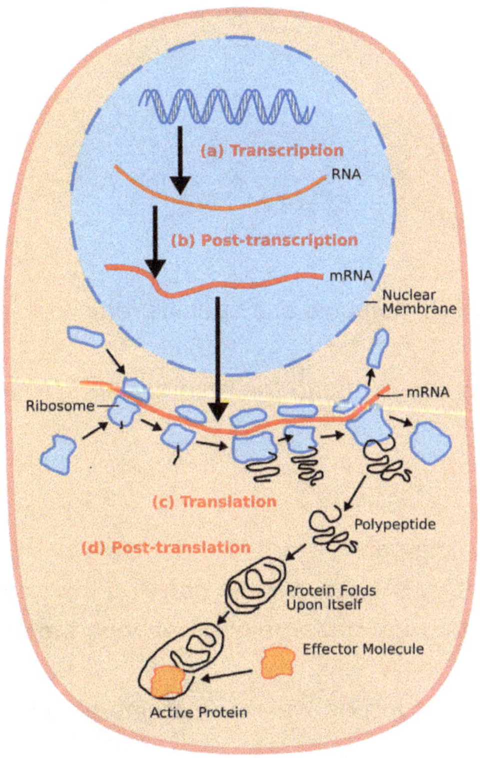

Ribosomes

Ribosomes are organelles that are mainly composed of a type of RNA called ribosomal RNA, or rRNA. All cells have ribosomes. Ribosomes are composed of two pieces, called subunits, aptly named the large and small subunits. The ribosome acts as a platform for the drama of translation to unfold and catalyzes the peptide bond formation between amino acids. The ribosome itself can't "read" the amino acids, and it needs another molecule to bring amino acids to the ribosome, as seen in Figure 9.10.

Figure 9.10. RNA Types

tRNA

If the ribosome can't "read" the mRNA, and if it needs another molecule to bring amino acids to the ribosome so peptide bonds can be built, what molecule can do this? That's where another type of RNA, called transfer RNA, or tRNA comes in. The tRNA is the molecule that decodes the mRNA into an amino acid, as seen in Figure 9.11.

How Polypeptides are Built

Initiation

- The small subunit of the ribosome binds to the mRNA molecule at a specific region called the start codon (usually AUG).
- The initiator tRNA, carrying the amino acid methionine, binds to the start codon.

Elongation

- The ribosome moves along the mRNA in a 5' to 3' direction.
- As each mRNA codon is exposed, a complementary tRNA molecule carrying the corresponding amino acid binds to the mRNA codon through base-pairing between the anticodon on tRNA and the codon on mRNA.
- The ribosome catalyzes the formation of a peptide bond between the amino acids carried by adjacent tRNA molecules.
- The ribosome then moves to the next codon, and the process is repeated. This step involves the continuous addition of amino acids to the growing polypeptide chain.

Termination

» The process continues until a stop codon (UAA, UAG, or UGA) is reached on the mRNA.
» There are no tRNA molecules with anticodons complementary to stop codons, so the ribosome machinery halts.
» The newly synthesized polypeptide chain is released from the ribosome.

Figure 9.11. Translation

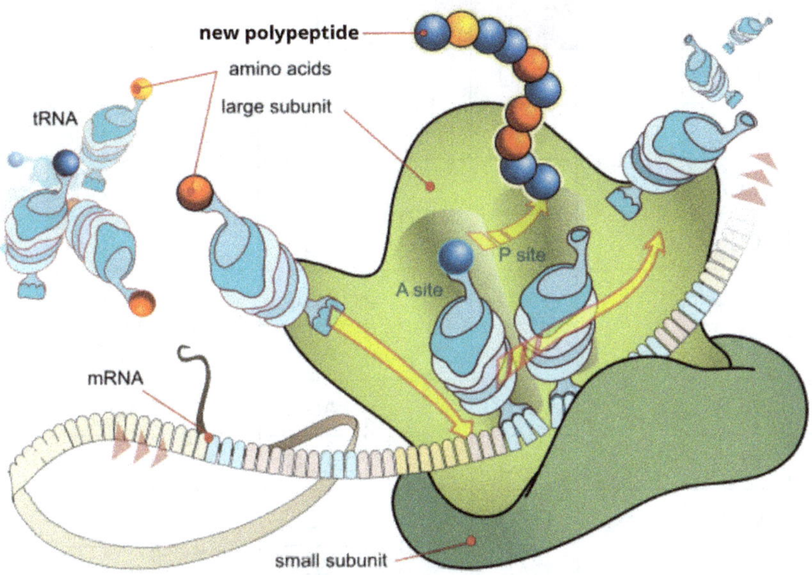

Practice Transcribing and Translating a Polypeptide

Example sequence:

» AUG CUC UUA UGG AGA UAC

Identify Codons:

» The mRNA sequence is read in sets of three nucleotides, known as codons. In this example: AUG, CUC, UUA, UGG, AGA, UAC

Refer to the Codon Chart:

» A codon chart (Figure 9.12), also known as a genetic code chart, associates each codon with the corresponding amino acid.

Figure 9.12. mRNA

First Base	Second Base: U		Second Base: C		Second Base: A		Second Base: G		Third Base
U	UUU	phe	UCU	ser	UAU	tyr	UGU	cys	U
	UUC		UCC		UAC		UGC		C
	UUA	leu	UCA		UAA	STOP	UGA	STOP	A
	UUG		UCG		UAG		UGG	trp	G
C	CUU		CCU	pro	CAU	his	CGU	arg	U
	CUC	leu	CCC		CAC		CGC		C
	CUA		CCA		CAA	gln	CGA		A
	CUG		CCG		CAG		CGG		G
A	AUU	ile	ACU	thr	AAU	asn	AGU	ser	U
	AUC		ACC		AAC		AGC		C
	AUA		ACA		AAA	lys	AGA	arg	A
	AUG		ACG		AAG		AGG		G
G	GUU	val	GCU	ala	GAU	asp	GGU	gly	U
	GUC		GCC		GAC		GGC		C
	GUA		GCA		GAA	glu	GGA		A
	GUG		GCG		GAG		GGG		G

9. Using the codon chart in Figure 9.7, translate each codon into its corresponding amino acid:

AUG: ______________________________

CUC: ______________________________

UUA: ______________________________

UGG: ______________________________

AGA: ______________________________

UAC: ______________________________

10. Besides RNA and amino acids, are other classes of molecules directly involved in the process of translation?

11. What are the different roles of mRNA, tRNA, and rRNA in protein synthesis?

12. Part of an intron was accidentally left in the mRNA—just one base! What would happen to the protein that was built from this mRNA? Give an example using the practice sequence above.

13. If the protein was to function within the cell, versus a protein that was to be secreted from the cell, would the location of translation differ? Explain.

Mutations

While DNA is a relatively stable molecule, mutations can occur which alter the sequence of DNA. A **mutation** is any change in the sequence - the A, T, C, and G's of the DNA molecule. Mutations can be small, single nucleotide substitutions, such as swapping an A where a G was, or they can be larger insertion or deletions of entire sequences of DNA. There are redundancies in the genetic code, for example see how many different codons are used to encode threonine (thr), so not all mutations cause changes to the structure of the protein.

Mutations can be silent mutations, as shown in Figure 9.13, meaning the change in the sequence of DNA does not change the protein that is created. Other mutations can cause a change in the protein that is created.

Figure 9.13. Silent Mutation

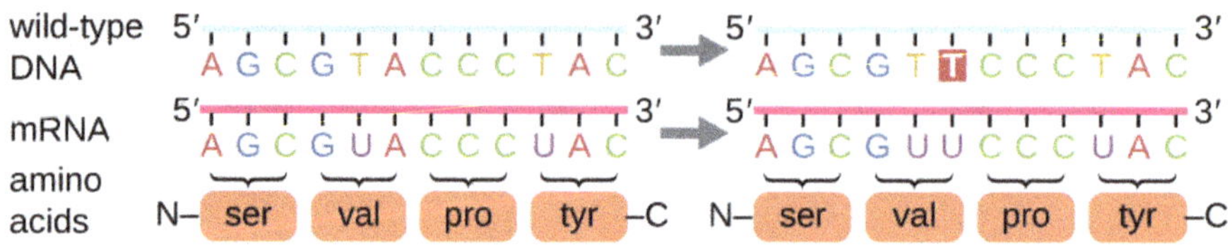

Mutations have a variety of causes. Sometimes there are errors during DNA replication that are not detected, and the resulting daughter cells carry a DNA sequence that is different from the parent cell. Sometimes environmental factors, like UV radiation or chemicals in the tar from tobacco, can damage the DNA molecule resulting in a change of the sequence. There are a wide variety of environmental and endogenous factors that can alter the sequence of DNA. However, not all changes to the sequence of DNA will result in harmful mutations. Mutations have the potential to change the shape of a protein, but whether that change in shape negatively or positively affects the function of the protein, or perhaps it doesn't affect it at all, depends on the type of mutation and how the protein has changed.

Let's go through some of the common types of mutations and their potential impacts:

9

» **Substitution**: This occurs when one base, let's say a T is swapped for another, like a G. Or, like in Figure 9.14, an A is substituted for a C. Depending on where in the coding sequence it happens, it could have an effect on the protein that is produced, or not. Substitution mutations are less likely to impact the protein structure when they occur on the third place of the codon due to the redundancy in the codon sequences. When only one base is substituted, we call this a point mutation.

Figure 9.14. Substitution Mutation

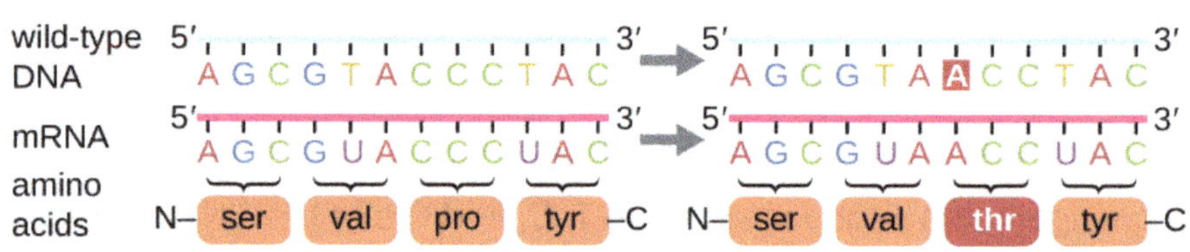

» **Frame Shift**: since the sequences of DNA are read in triplet codons (each three bases corresponding to one amino acid) changing the reading frame of the codons by inserting or deleting one or more bases from the sequence will usually have drastic impacts on the protein which is produced. This is because inserting or deleting bases that are not sets of three shifts how the entire sequence is read from the point of the change and everything downstream. Figure 9.15 shows a frameshift mutation caused by deleting two bases.

- **Insertion**: one or more bases is added to the DNA sequence in a coding region, altering the entire sequence downstream of the insertion.
- **Deletion**: one or more bases is removed from the DNA sequence in a coding region, altering the entire sequence downstream of the deletion.

Figure 9.15. Frame Shift Mutation

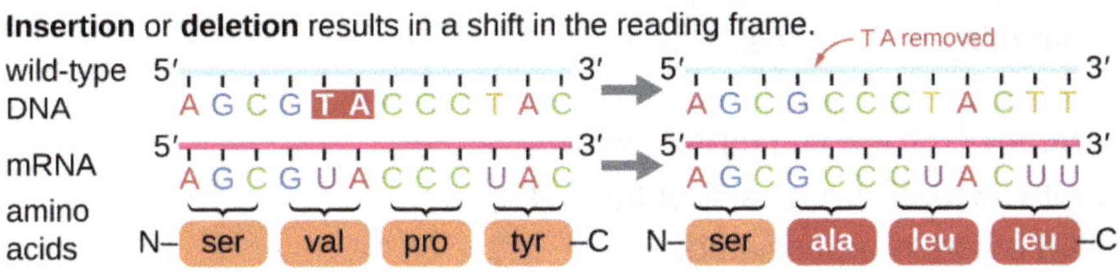

14. Why are insertion or deletion mutations more likely to cause a change in the structure of the protein than a substitution mutation?

Multiple Choice Review

1. What is the main function of mRNA in the process of transcription and translation?
   a. To replicate DNA
   b. To carry genetic information from DNA to the ribosome
   c. To bring amino acids to the ribosome
   d. To make a copy of itself

2. During translation, what role do ribosomes play?
   a. They replicate DNA.
   b. They synthesize proteins by reading mRNA sequences.
   c. They transport proteins to different parts of the cell.
   d. They modify mRNA before it is translated.

3. Which molecule is responsible for bringing the correct amino acid to the ribosome during translation?
   a. mRNA
   b. rRNA
   c. tRNA
   d. DNA

4. In the context of transcription, what is the significance of the promoter region in DNA?
   a. It signals the end of a gene.
   b. It marks the spot where replication begins.
   c. It is the binding site for RNA polymerase to start transcription.
   d. It is the binding site for ribosomes to start translation.

5. How does the process of translation terminate?
   a. When RNA polymerase detaches from the DNA
   b. When a stop codon is reached on the mRNA strand
   c. When the ribosome reaches the end of the mRNA strand
   d. When all amino acids have been used up

6. If a mutation occurs in the promoter region of a gene that prevents the binding of RNA polymerase, what would be the immediate impact on gene expression?
   a. Increased transcription of the gene
   b. Unchanged transcription of the gene
   c. Decreased transcription of the gene
   d. Increased translation of the mRNA

7. A scientist introduces a nucleotide sequence into a cell that results in the production of a premature stop codon in the mRNA. What is the likely outcome for the protein encoded by this mRNA?
   a. The protein will be longer than normal.
   b. The protein will be shorter than normal.
   c. The protein will have an altered function but normal length.
   d. The protein's production will not be affected.

8. Considering the process of translation, what would be the consequence if a tRNA molecule carrying an amino acid had an anticodon that was complementary to a stop codon?
   a. Translation would not initiate.
   b. Amino acids would not be brought to the ribosome.
   c. The protein synthesis would prematurely terminate.
   d. The protein synthesis would extend beyond the normal stop signal.

9. In a laboratory experiment, a researcher observes that after transcription, the mRNA transcript contains sequences that are not present in the mature mRNA. What process explains the removal of these sequences?
   a. Replication
   b. RNA editing
   c. Splicing
   d. Polyadenylation

10. If a mutation changes a codon in the mRNA to one that codes for the same amino acid, what is the effect on the protein produced?
    a. The protein will have a different primary structure.
    b. The protein will be synthesized faster.
    c. The protein will be synthesized slower.
    d. The protein will remain unchanged.

10.1 Aerobic Cellular Respiration

Cellular respiration is the process in which the bonds in food molecules are broken to release energy. The energy in the bonds of food molecules is used to produce adenosine triphosphate (ATP). ATP is a usable form of energy for all living organisms. So, to understand cellular respiration, let's review chemical bonds. Figure 8.1 is a diagram of the structural model of glucose. While many molecules can be used for cellular respiration, we will focus on how glucose is used in cellular respiration.

Figure 8.1. Structural Model of Glucose

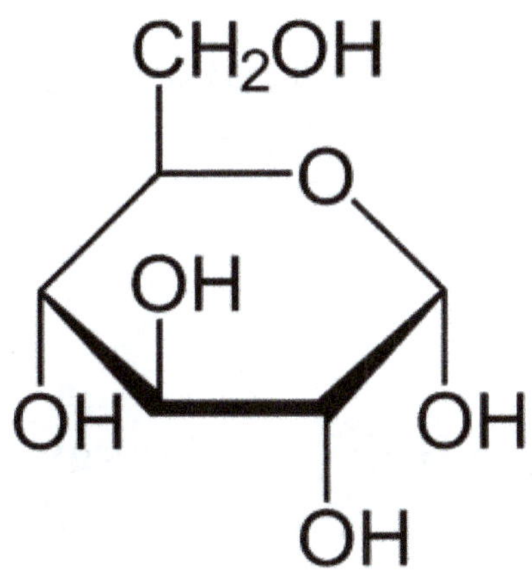

1. What type of bond occurs between the atoms in glucose?

2. What is being shared between the atoms in glucose?

10

3. What type of biological macromolecule is glucose?

Redox Reactions

Redox reactions are chemical reactions that move electrons around. In redox reactions, one molecule loses electrons, and one molecule gains those electrons. When a molecule loses electrons, it has been oxidized. When one molecule gains electrons, it has been reduced. Redox reactions always happen in pairs since the electrons are transferred from one molecule to another. The mnemonic "LEO says GER" helps to remember which is which: Gain electrons, reduction (GER); lose electrons, oxidation (LEO).

During cellular respiration, glucose is oxidized and forms CO_2. The electrons taken off glucose eventually end up being added to an oxygen molecule (O_2), but they have quite the journey to make with a few stops before they do, as seen in Figure 8.2.

Figure 8.2. Cellular Respiration Reaction

becomes oxidized ($C_6H_{12}O_6$ → $6\,CO_2$)

$$C_6H_{12}O_6 + 6\,O_2 \longrightarrow 6\,CO_2 + 6\,H_2O + \text{Energy}$$

becomes reduced ($6\,O_2$ → $6\,H_2O$)

4. What is removed during oxidation?

5. In redox reactions, electrons are ____________ from the molecule that is ___________ and transferred to the molecule that gets ________________________.

6. What molecule is oxidized during cellular respiration? What does it become after it's oxidized?

7. What molecule is reduced during cellular respiration? What does it become after it's reduced?

What is ATP?

Adenosine triphosphate, or ATP, is the main energy source for cells. It is made up of an adenosine nucleoside with three phosphate groups. It takes a lot of energy to get three negatively charged phosphate groups to bond together in a row like that. Energy can be released from ATP by breaking off the third phosphate group, which forms ADP. ADP can be recycled into ATP, it just requires an input of energy and a phosphate group to "recharge" ADP into ATP, as seen in Figure 8.3.

Figure 8.3. Adenosine Triphosphate (ATP)

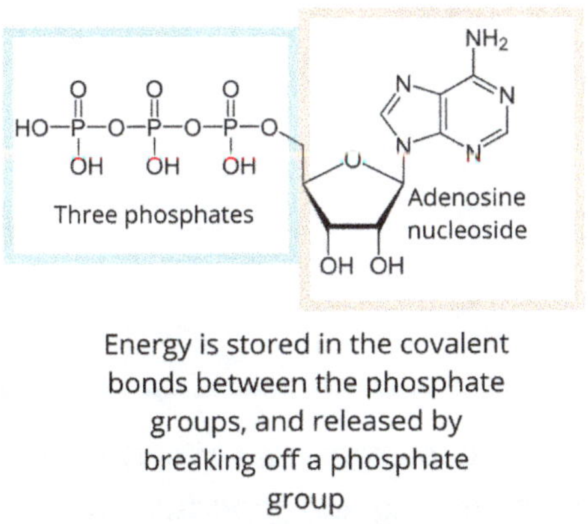

ATP is the energy that cells use to do basically everything within the cell that requires energy, as seen in Figure 8.4. This includes active transport, cell signaling, reproduction, and generally maintaining homeostasis. And when it comes down to it, no homeostasis, no life!

Figure 8.4. ATP Cycling

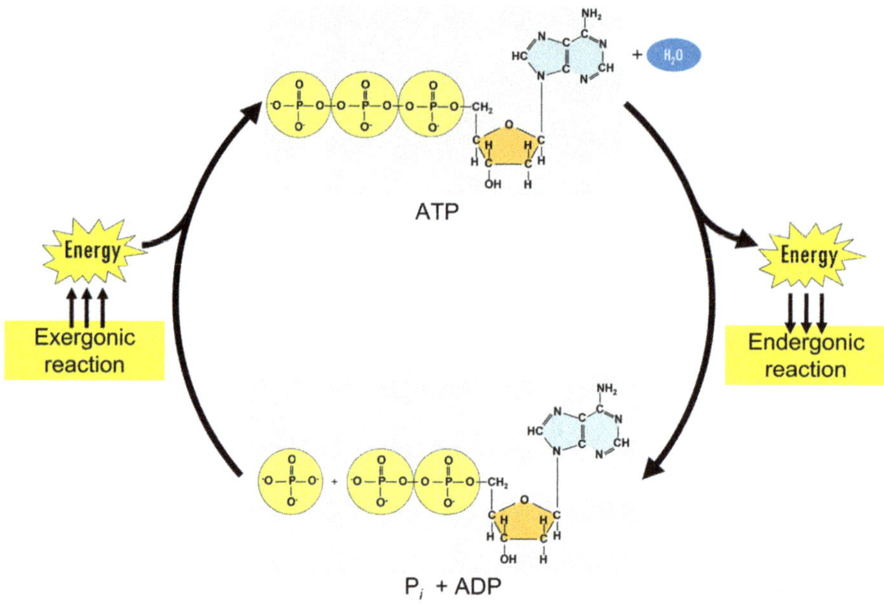

8. What do cells use ATP for?

9. What has more energy: ADP or ATP?

Mitochondria

Mitochondria are often shown as weird little pinto-bean-looking organelles in cartoon drawings of cells, like in Figure 8.5. But they're one of the most interesting and important organelles within eukaryotic cells. Before they became organelles through the process of endosymbiosis, mitochondria were free-living prokaryotes. So, they have their own DNA and ribosomes and reproduce on their own. We'll be tracing where the different steps of cellular respiration occur, so it helps to know some mitochondrial anatomy.

Figure 8.5. The Mitochondria

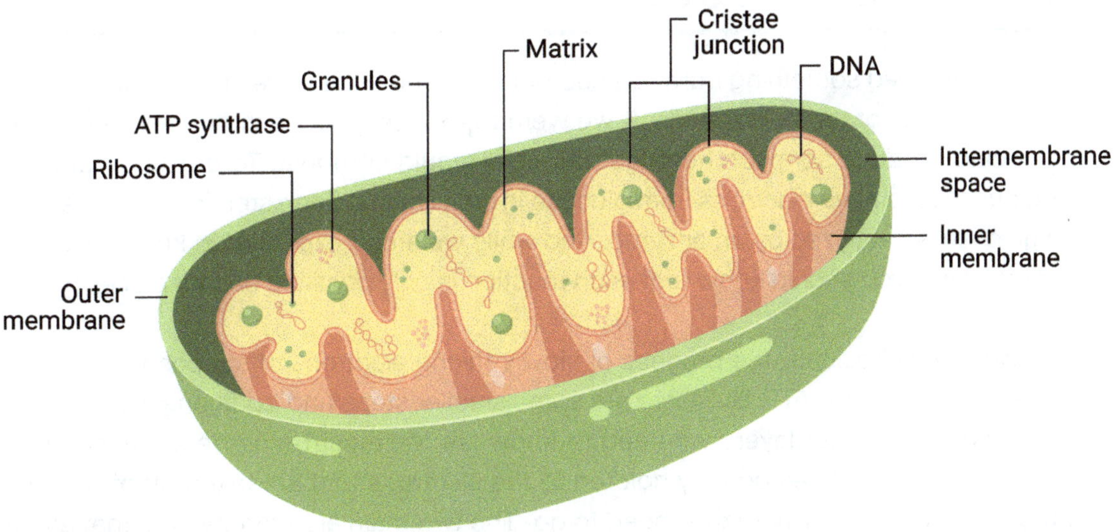

There are a few important things to know about the mitochondria. First, it has two membranes, an outer and an inner membrane. There is a space between these membranes, called the intermembrane space, which will be important for understanding how the electron transport chain works.

The inner membrane is folded up on itself, back and forth, to increase the surface area of this membrane. We call the folds of the inner membrane cristae.

The interior of the mitochondria is called the matrix. We can think of the matrix as the cytoplasm of the prokaryotic cell that the mitochondria used to be. It's a fluid-filled area full of enzymes and other molecules dissolved in water. This area is important because this is where the citric acid cycle occurs.

10. How did mitochondria become organelles?

11. What are the important features/regions of mitochondria that we need to know about for cellular respiration?

Electron Carriers

Have you ever watched something burn? Zoned out watching a fire? Fire, the release of heat and light, is the result of redox reactions. If we were to oxidize glucose in our cells all in one step, that would produce so much heat that our cells would combust. To prevent that from happening our cells remove electrons from glucose in small steps. Each step in this process of oxidizing glucose is catalyzed by an enzyme. And while we aren't expected to know these enzymes for our class, you should know that each reaction is only possible because of enzymes.

Cellular respiration is basically the process of slowly taking electrons off glucose in order to use those electrons to build ATP. But there are some smaller steps and additional players that we need to know. The first players we need to know are the electron carriers. These molecules are the ones that will temporarily hold on to the electrons that are taken off of glucose and shuttle the electrons to where they need to go. The two main electron carriers that are involved in cellular respiration are NAD^+ and FAD. NAD^+ is the oxidized form of NADH. So, when NAD^+ has gained some electrons from glucose, it becomes NADH (Figure 8.6).

Figure 8.6. Electron Carrier Nicotinamide Adenine Dinucleotide (NAD) + Hydrogen (H), NADH

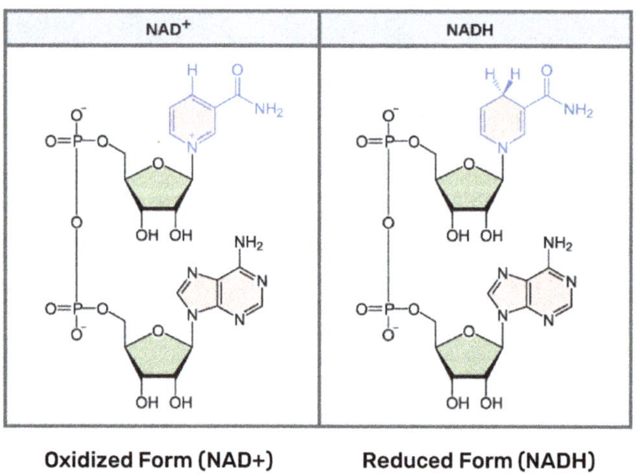

FAD is the oxidized form of $FADH_2$. When FAD has gained some electrons from glucose, it becomes $FADH_2$ as seen in Figure 8.7.

Figure 8.7. Electron Carrier Flavin Adenide Dinucleotide or $FADH_2$

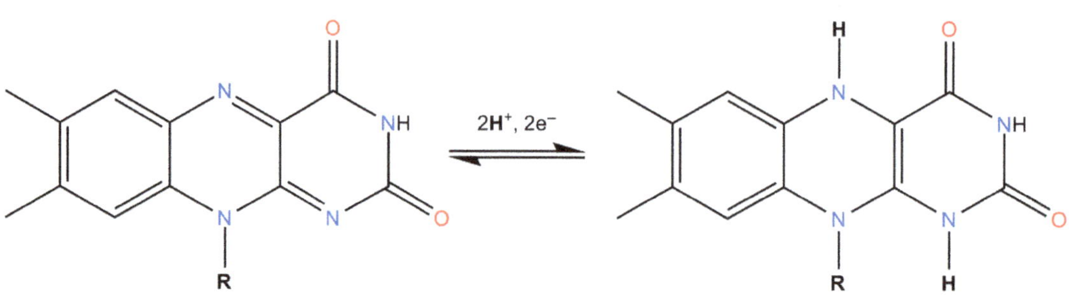

These electron carriers are akin to your paycheck when it's in the mail. Your money is there, it's technically yours, but it's in limbo until it's been delivered to your house, and you cash the check. While the electrons from glucose are attached to NADH or $FADH_2$, they don't represent a useful form of energy, but they will help get those electrons to where they need to go in order to build ATP, as seen in Figure 8.8.

Figure 8.8. NADH Reduction

$$\mathbf{NAD^+ + 2e^- \rightarrow NADH}$$

12. What is the function of NADH and $FADH_2$?

13. What is the oxidized form of NADH and $FADH_2$? What is its reduced form?

14. What are NADH and $FADH_2$ carrying?

10

Electrons and Cellular Respiration

During cellular respiration, electrons are removed from glucose. Oxidation of glucose results in the production of carbon dioxide. This means the carbon dioxide you exhale is actually the remnant of carbon from food that you ate. The cellular respiration reaction is shown in Figure 8.9.

Figure 8.9. Cellular Respiration Reaction

$$\underbrace{C_6H_{12}O_6}_{\text{becomes oxidized} \to 6\,CO_2} + \underbrace{6\,O_2}_{\text{becomes reduced} \to 6\,H_2O} \longrightarrow 6\,CO_2 + 6\,H_2O + \text{Energy}$$

Glucose is oxidized in many small steps, each catalyzed by an enzyme as seen in Figure 8.10. The electrons removed from glucose represent a form of energy, and these electrons will be temporarily stored with the electron carriers NADH and $FADH_2$. The first three steps of cellular respiration—glycolysis, pyruvate oxidation, and the citric acid cycle—are responsible for the oxidation of glucose to carbon dioxide and the transfer of electrons from glucose to the electron carriers.

Figure 8.10. Movement of Electrons and Energy in Cellular Respiration

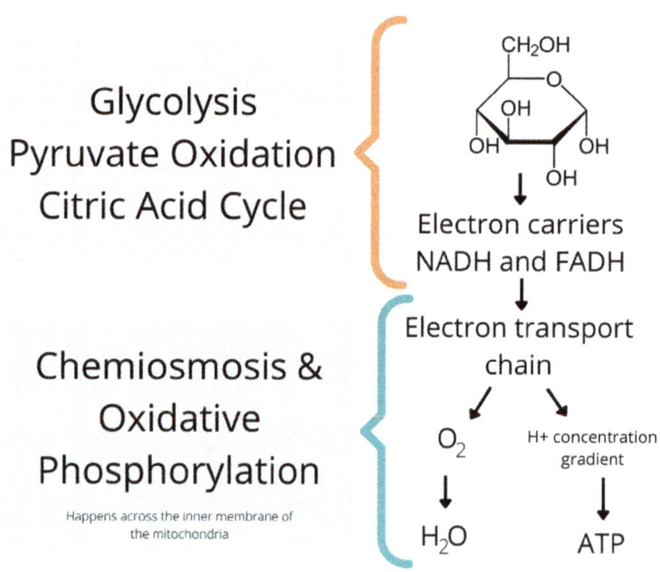

Once glycolysis, pyruvate oxidation, and the citric acid cycle have completed the oxidation of glucose, the electrons being held by NADH and $FADH_2$ will then make their way to the inner membrane of the mitochondria. Here, the energy of the electrons that were harvested from glucose will be used to create a **concentration gradient** of H^+ ions across the inner membrane of the mitochondria by the enzymes of the **electron transport chain**. This concentration gradient provides the energy necessary for the enzyme **ATP synthase** to build ATP from ADP (adenosine diphosphate) and a free phosphate group.

But what about oxygen? Where does it come in? Remember those enzymes in the electron transport chain that used the electrons to create a Hydrogen ion concentration gradient? After those enzymes have used the energy in the electrons they were given to move H^+ against its concentration gradient the electrons are relatively low energy, like a dead battery. Since those electrons are not useful anymore, they will get transferred to O_2, which splits apart and bonds with some free-floating H^+ ions to form two water molecules. This is why O_2 is called the **final electron acceptor** in cellular respiration, as it's the last stop electrons make in their journey.

15. What molecule is oxidized during cellular respiration? What does it form after oxidation of this molecule has been completed?

16. What molecule is reduced during cellular respiration? What does it form once it is reduced?

17. Do the electrons removed during the oxidation of glucose get transferred directly to ATP?

18. How do the electrons get transferred to the electron transport chain?

10

19. The electron transport chain uses electrons for active transport. What are they moving across the inner membrane of the mitochondria, and in what direction?

20. Why is O_2 called the final electron acceptor?

Overview of Cellular Respiration

You should be able to recreate the diagram in Figure 8.11 and explain what is happening at each step.

Figure 8.11. Overview of Cellular Respiration Steps

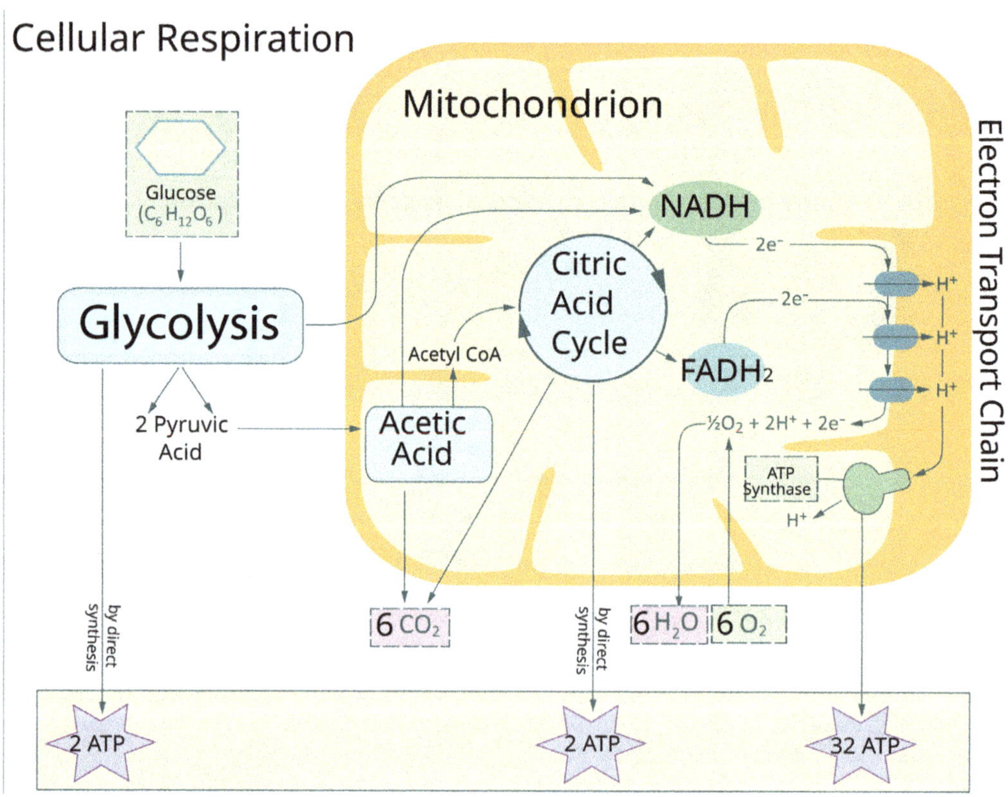

Cellular Respiration Process

In this section, we will dive into the main steps of cellular respiration in more detail. For this class, it is not expected to know the enzymes involved in each step. However, you should know the reactants and products of each step, and where it occurs. Additionally, you should be able to determine what would happen if one piece of this process was missing or non-functional.

Glycolysis

The first step of cellular respiration occurs in the cytoplasm. All cells, even those without mitochondria, can complete glycolysis. Additionally, this step does not require oxygen, so it can occur in anaerobic (oxygen-lacking) environments to produce a small amount of ATP.

Glycolysis has two main steps: the energy investment step and the energy harvesting step. First, two phosphate groups are removed from ATP and attached to glucose. This results in two molecules of ADP and a molecule known as fructose disphosphate. Fructose disphosphate (a 6-carbon molecule) quickly breaks apart into two molecules of glyceraldehyde 3-phosphate (3 carbon molecules). So, we've used some energy to break apart glucose (6 carbons) into two three-carbon molecules, as seen in Figure 8.12.

Figure 8.12. Glycolysis

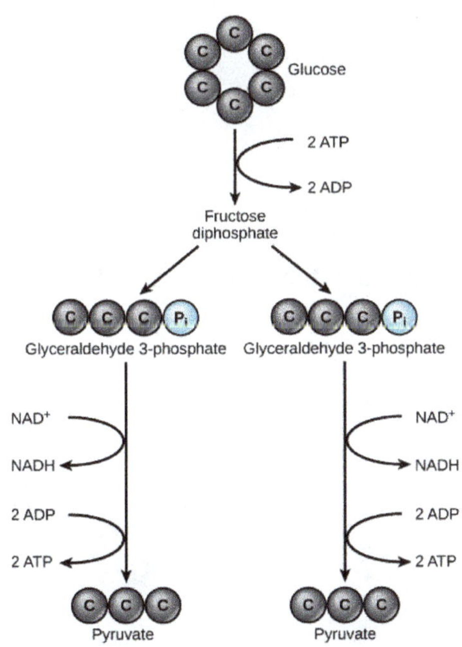

The main purpose of the first three steps of cellular respiration is to remove electrons from glucose (to oxidize glucose). This happens in the second step of glycolysis. Each glyceraldehyde 3-phosphate molecule is oxidized. Two electrons are removed and transferred to NAD^+

to form NADH. Then, the phosphate groups are also removed and some of the energy that was invested is recouped, plus a little extra. Each glyceraldehyde 3-phosphate molecule is oxidized again to form 2 ATP. Since we have two glyceraldehyde 3-phosphate, that means we make a total of 4 ATP and 2 NADH by the end of glycolysis. Glyceraldehyde 3-phosphate is oxidized to pyruvate (sometimes called pyruvic acid). Pyruvate is the molecule that we will follow into the next step.

21. What do you think the term glycolysis means? Consider the prefix *glyco* and the suffix -*lysis*.

22. What does the name glycolysis have to do with what happens during this step?

23. Where does glycolysis occur within the cell?

10

24. Does glycolysis require mitochondria or oxygen? What does this tell you about the types of organisms that can go through glycolysis?

25. What are the reactants of glycolysis?

26. What are the products of glycolysis?

27. How much net ATP is made during glycolysis? Remember, net is what is left over after accounting for anything that was invested.

28. What molecule is reduced during glycolysis that will be used later during the electron transport chain?

10

Pyruvate Oxidation

When oxygen is available, the pyruvate molecule will be transported into the mitochondria. As this is happening, the three-carbon pyruvate will be oxidized to form a two-carbon molecule (acetyl). The carbon that is removed is exhaled as carbon dioxide, a waste product. A coenzyme molecule, called coenzyme A, is attached to the two remaining carbons to form the molecule called acetyl CoA. The process of oxidizing pyruvate to form acetyl CoA also reduces NAD^+ to NADH as seen in Figure 8.13.

Figure 8.13. Pyruvate Oxidation

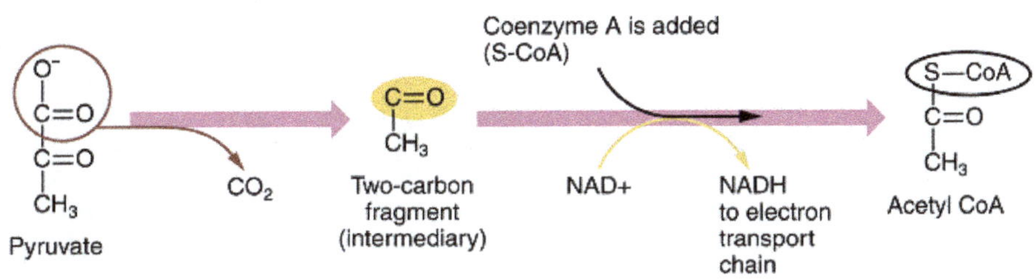

The carbons that were originally in glucose have now been converted into two carbon dioxide molecules and two acetyl CoA molecules. Remember, glycolysis ends with two pyruvates, and both pyruvates will be converted to acetyl CoA and carbon dioxide in this step.

Now, our Acetyl CoA molecule is in the matrix of the mitochondria. The coenzyme A is a coenzyme that is necessary to activate the enzymes of the next step, the citric acid cycle. Acetyl CoA has two purposes: it delivers the remaining carbons from glucose to the citric acid cycle and activates the enzymes of the citric acid cycle.

29. Where does pyruvate oxidation occur?

30. What are the products of this step?

31. What are the reactants of this step?

10

32. How many carbons are in pyruvate? How many in acetyl CoA? Why don't these numbers match? What happened to the missing carbon?

33. What other important molecule is reduced during pyruvate oxidation? Where will this molecule eventually end up?

34. Some texts call this step the "linking step." Why would it be called this?

35. What is a coenzyme? What is coenzyme A used for?

36. Where does Acetyl CoA go next?

10

Citric Acid Cycle

We ended the last step in the matrix of the mitochondria. The acetyl CoA molecule will now move through the citric acid cycle in the matrix of the mitochondria. The citric acid cycle is sometimes also called the Krebs cycle, those terms are synonymous.

This step completes the oxidation of what was glucose and the only thing left is carbon dioxide. The main purpose of the citric acid cycle is to remove as many electrons from acetyl CoA as possible and transfer those electrons to NAD^+ and FAD, as seen in Figure 8.14.

Figure 8.14. The Citric Acid Cycle

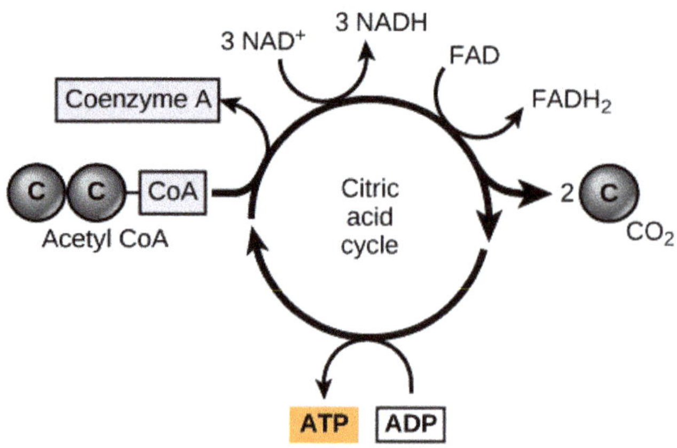

The citric acid cycle also produces one molecule of ATP for each acetyl-CoA molecule. Again, for our purposes, it's not essential to know the individual steps or enzymes of the citric acid cycle. You should know the reactants, the products, and where it occurs.

37. Where does the citric acid cycle take place?

38. Do you think prokaryotes can go through the citric acid cycle? Why or why not?

39. What is the main purpose of the citric acid cycle?

40. By the end of this step, the energy that was contained in the covalent bonds of glucose in the form of electrons has been transferred to what molecules?

41. What waste products are created during the citric acid cycle? What happens to this molecule?

Electron Transport Chain

The electron transport chain (ETC) is the main ATP-generating step of aerobic cellular respiration. The first three steps of cellular respiration have been harvesting electrons from glucose. Here's where those electrons will be used to generate the energy needed to build ATP. But, those electrons still have quite a journey to make. And remember, the electrons from glucose don't end up as part of the ATP molecule. They provide an energy source that can be used to build ATP.

Remember NADH and $FADH_2$? This step, the electron transport chain, is where these electron carriers end up. Their whole job is to carry electrons to enzymes embedded within the cristae of the mitochondria. This collection of enzymes spanning the inner membrane of the mitochondria is what we call the electron transport chain. NADH and $FADH_2$ are oxidized at the ETC, forming NAD^+ and FAD. Once they're oxidized, they will return to the other steps of cellular respiration and be reused.

Before we dive into how this process works, let's review active transport. At times, cells will need to move molecules so they are concentrated on one side of a membrane. Moving molecules against their concentration gradient, or from low to high concentration, takes energy. When we talked about active transport previously, we said cells use ATP as the energy source to power active transport. The active transport we'll see happening in the electron transport chain doesn't use ATP, though. Here, these enzymes use electrons as an energy source to carry out active transport. Electrons moving is basically electricity. These enzymes are, in a way, using electricity (in the form of electrons) to move molecules from a low concentration to a high concentration, as seen in Figure 8.15.

Figure 8.15. The Electron Transport Chain, Chemiosmosis, and ATP Synthase

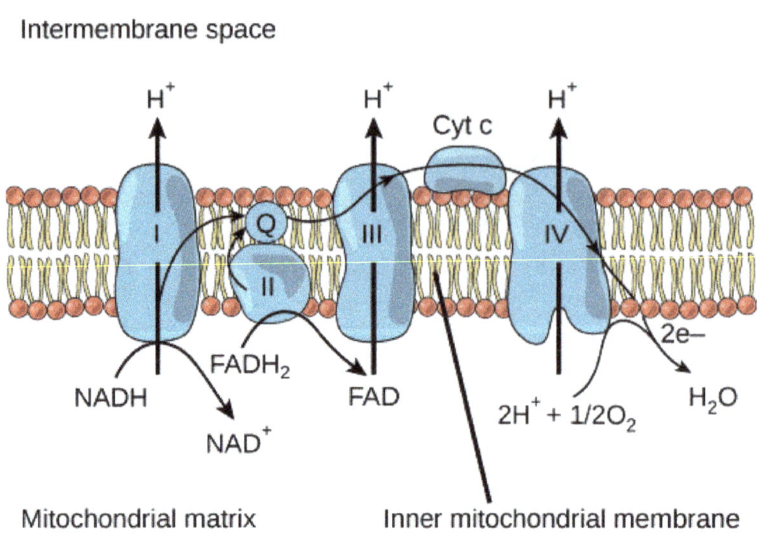

But what are these enzymes in the electron transport chain using the electrons from NADH and $FADH_2$ to move? They're moving Hydrogen ions. More specifically, they're moving H^+ from a low-concentration area in the matrix to a high-concentration area in the intermembrane space. Students will often ask why H^+? There's no real good answer, besides that's how life evolved, and it's likely because H^+ ions are common.

The enzymes in the electron transport chain use the energy from the electrons that were taken off of glucose to create a high concentration of H^+ ions in the intermembrane space. Once NADH or $FADH_2$ gives its electrons to the ETC, those electrons are used to pump H^+, then the electrons are passed to the next enzyme, which uses the electrons as an energy source to pump H^+. By the time the electrons have made it to the end of the ETC, they're relatively low in energy, so they aren't useful anymore. These low-energy electrons are passed off to an O_2 molecule. This causes O_2 to split apart and form two O^{2-} ions. The O^{2-} ions bond to two H^+ ions to form water. This serves two purposes: first, this prevents the final enzyme in the chain from becoming "full" of electrons and keeps the chain moving. Second, this helps to reinforce the concentration gradient of H^+ the ETC is creating by removing more H^+ ions from the matrix. Isn't this amazing? This is where the oxygen we inhale goes, this is what it's used for. It turns into water at the end of the ETC.

42. Would the pH of the intermembrane space be acidic or basic? Why?

43. Is the movement of H^+ ions at the ETC considered active transport? Explain.

44. Think about what types of molecules can move directly through a membrane (without a protein). Why don't the H^+ ions that are pumped into the intermembrane space just move back down into the matrix, in the direction that they would normally diffuse?

45. What role does oxygen play in cellular respiration? What would happen if oxygen wasn't available?

46. If NADH and $FADH_2$ can't unload their electrons at the ETC, what would happen to the other steps of cellular respiration?

We left off with the ETC having created a concentration gradient of H^+. The electrons from glucose are used to concentrate H^+ in the intermembrane space of the mitochondria. But why? Isn't the whole purpose of all this to make ATP? Where's the ATP? Hold onto your hydrogens, here's where we make ATP! Those H^+ ions that are concentrated in the intermembrane space are a form of potential energy, kind of like water behind a dam. Those H^+ ions "want" to diffuse: they want to move across the membrane and diffuse until they reach equilibrium. But because they're ions, they can't move through the plasma membrane directly, they require a protein to transport them across. In the mitochondria, these H^+ ions only have one protein they can move through in order to move down their concentration gradient: ATP synthase. This enzyme provides a passage for H^+ ions to move back to a low-concentration area (the matrix). While letting these H^+ ions move down their concentration gradient, ATP synthase taps into the potential energy that is released as H^+ moves down its concentration gradient. It's this energy that powers the production of the high-energy bond between ADP and a third phosphate group, as seen in Figure 8.16.

Figure 8.16. The ATP Synthase

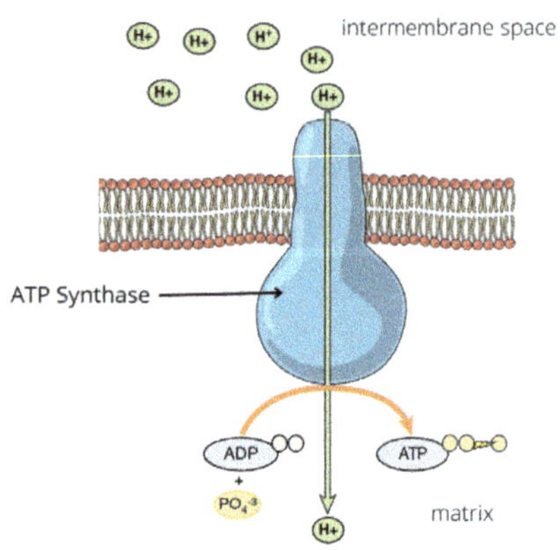

Let's continue with the dam analogy. We can make electricity by letting water flow through a dam, and as the water flows, it turns a turbine. This turning of the turbine generates electricity. ATP synthase is like the turbine: as H^+ moves through ATP synthase, it powers the reaction that builds ATP. ATP synthase even turns like a little turbine as H^+ moves through.

47. What type of biological macromolecule is ATP synthase?

48. What is the energy source ATP synthase uses to build ATP?

49. The inner membrane of the mitochondria suddenly became permeable to H^+ ions. What would happen to ATP production?

10

50. How is ATP synthase similar to a hydroelectric dam?

Aerobic Cellular Respiration Review

1. In what part of the cell does glycolysis take place?

2. List the reactants and products of glycolysis.

3. Describe the event that occurs when pyruvate enters the mitochondria?

4. In what major area of the mitochondria does the Citric Acid Cycle take place?

5. List the reactants and products of the Citric Acid Cycle.

6. Identify the major structure or area where the Electron Transport Chain takes place?

7. What subatomic particle is transferred to the Electron Transport Chain from the Citric Acid Cycle? What molecule(s) carries this subatomic particle?

8. What reactant molecule from outside the cell is essential for electron transport to occur? Describe the part of the diagram you focused on to answer this question.

9. What molecule is the final acceptor of the electrons from the ETC?

10. Describe two main functions of the ETC.

11. What steps of cellular respiration are catabolic? Anabolic?

12. Cellular respiration breaks down glucose and releases carbon dioxide and water. What molecule is the source of the carbon dioxide? What molecule is the source of the water?

13. What steps of cellular respiration oxidize glucose? What is reduced during these steps?

14. The electrons from glucose are used to create ______________________ which provides the energy for ______________________ to build ______________________.

10

Multiple Choice Review

1. What is the main purpose of aerobic cellular respiration?
   a. To produce carbon dioxide and water
   b. To generate ATP for cellular energy
   c. To consume ATP
   d. To produce oxygen

2. Where does the Krebs cycle (citric acid cycle) occur within a eukaryotic cell?
   a. In the cytoplasm
   b. Within the mitochondrial matrix
   c. Across the inner mitochondrial membrane
   d. In the nucleus

3. Which of the following is a product of the electron transport chain in aerobic respiration?
   a. Glucose
   b. Pyruvate
   c. ATP
   d. Lactic acid

4. During aerobic respiration, oxygen is used at which stage?
   a. Glycolysis
   b. Citric Acid Cycle
   c. Electron transport chain
   d. Fermentation

5. What happens to the energy stored in glucose during aerobic cellular respiration?
   a. It is all converted into heat.
   b. It is stored in the form of ATP.
   c. It is entirely converted into carbon dioxide.
   d. It is used to produce oxygen.

6. If a cell's mitochondria are damaged, limiting its ability to perform the electron transport chain, how would this affect the cell's overall production of ATP during aerobic respiration?
   a. ATP production would increase due to higher glycolysis activity.
   b. ATP production would decrease due to impaired oxidative phosphorylation.
   c. ATP production would remain unchanged as glycolysis compensates.

7. If a person ingests a substance that acts as a competitive inhibitor for an enzyme in the citric acid cycle, what would be the immediate effect on aerobic respiration?
   a. ATP production would shift entirely to the citric acid cycle.
   b. An increase in acetyl-CoA production
   c. A decrease in ATP production due to slowed citric acid cycle reactions.
   d. An increase in ATP production as alternative pathways are utilized.
   e. No change in ATP production as the electron transport chain compensates.

10

8. How would the introduction of an artificial electron acceptor with a higher affinity for electrons than oxygen affect aerobic respiration?
   a. It would increase the efficiency of ATP production by accelerating the electron transport chain.
   b. It would decrease ATP production as it disrupts the proton gradient formation.
   c. It would have no effect on ATP production rates.
   d. It would cause the electron transport chain to stall, reducing ATP production.

9. In an experiment, a biologist finds that lowering the oxygen concentration around aerobic bacteria (bacteria that can do cellular respiration) slows their growth rate. What is the most likely explanation for this observation?
   a. The reduced oxygen levels limit glycolysis, reducing ATP production.
   b. The lower oxygen levels decrease the efficiency of the electron transport chain, reducing ATP production.
   c. The reduced oxygen concentration increases the citric acid cycle activity, but ATP production is unaffected.
   d. The lower oxygen levels enhance anaerobic pathways, which are less efficient at producing ATP.

Appendix A. Periodic Table of Chemical Elements

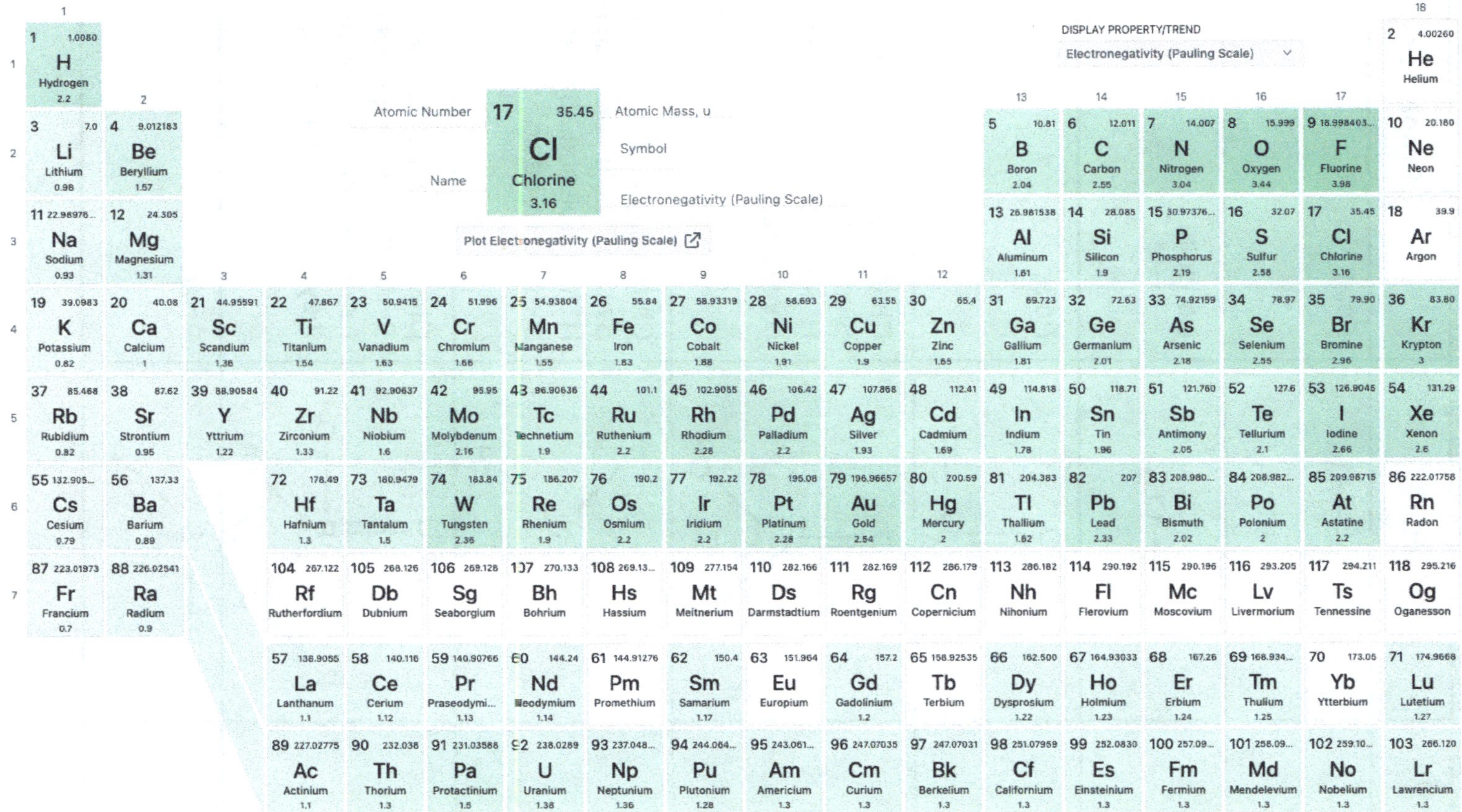

A

Appendix B. Common Organic Functional Groups

Functional Group	Structure	Properties
Hydroxyl	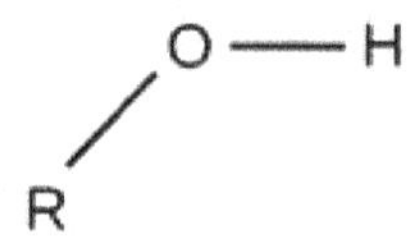	Polar
Methyl	$R-CH_3$	Nonpolar
Carbonyl	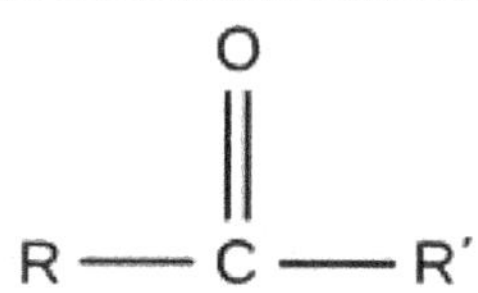	Polar
Carboxyl	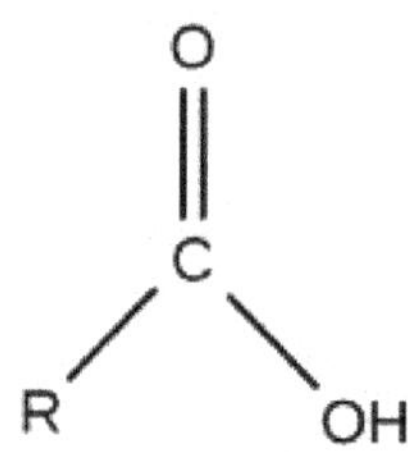	Charged, ionizes to release H^+. Since carboxyl groups can release H^+ ions into solution, they are considered acidic.
Amino	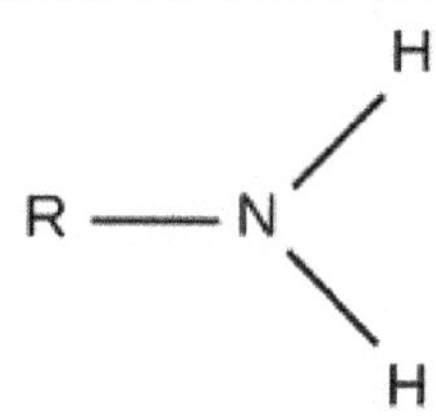	Charged, accepts H^+ to form NH_3^+. Since amino groups can remove H^+ from solution, they are considered basic.
Phosphate	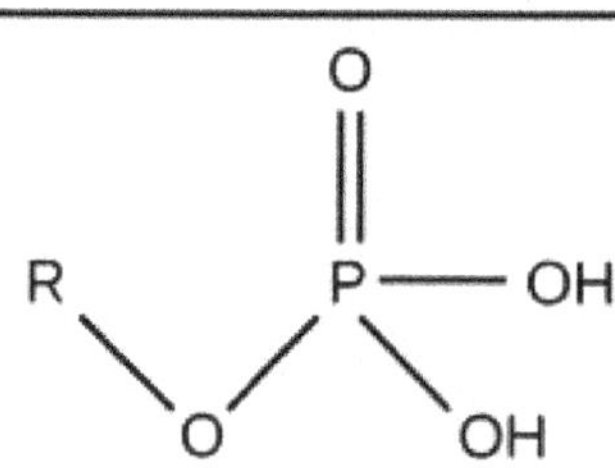	Charged, ionizes to release H^+. Since phosphate groups can release H^+ ions into solution, they are considered acidic.
Sulfhydryl	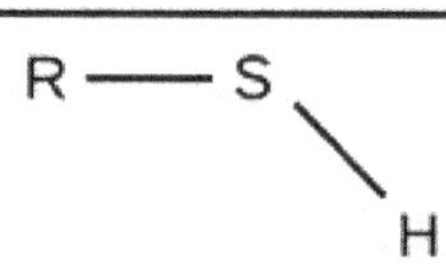	Polar

Appendix C. Glossary

5' G cap: a modified guanine nucleotide added to the 5' end of eukaryotic mRNA, crucial for its stability, nuclear export, and translation. (pg. 166)

acidic: a solution that has a pH lower than 7. (pg. 65)

active transport: a type of membrane transport that requires the cell to expend energy. (pg. 118)

adhesion: the tendency of water to make hydrogen bonds to other polar substances. (pg. 60)

allele: alternate versions of a gene that play a role in dominant and recessive genetic patterns. (pg. 150)

apoptosis: programmed cell death. (pg. 145)

ATP synthase: an enzyme that builds ATP from ADP and a free phosphate group. (pg. 185)

basic: a solution that has a pH higher than 7. (pg. 65)

buffers: molecules that minimize changes in pH that are caused by normal everyday activities of organisms and cells by bonding or donating Hydrogen ions. (pg. 71)

capillary action: the combined force of attraction among water molecules and the molecules of surrounding materials, causing a liquid to climb upward against the force of gravity. (pg. 60)

cellulose: a polysaccharide found in plants and is the main structural component of cell walls. (pg. 86)

cohesion: the tendency for water molecules to form weak hydrogren bonds and stick to each other. (pg. 60)

colloid: a type of heterogeneous mixture where small particles of one substance are dispersed throughout another substance, typically a liquid, but the particles are small enough that they do not settle out over time or are not visible to the naked eye. (pg. 35)

compound: a pure substance that is made up of two or more different types of atoms that are chemically combined in a fixed ratio. (pg. 32)

concentration gradient: the differential distribution of a substance across a space, driving its passive movement from high to low concentration, providing the energy necessary for the enzyme ATP synthase to build ATP from ADP (adenosine diphosphate) and a free phosphate group. (pg. 184)

control group: a group exposed to the same experimental conditions as the treatment group but does not receive the experimental treatment (independent variable). (pg. 17)

controlled variable: factors in an experiment that are intentionally kept constant or controlled to prevent them from influencing the results. (pg. 17)

crossing over: part of meiosis where homologous chromosomes exchange genetic information. (pg. 146)

dehydration synthesis: a process where two molecules combine to form a larger molecule, with the removal of water. (pg. 77)

dependent variable: the variable that is measured or observed to determine the effects of the independent variable; the outcome or response that researchers are interested in studying. (pg. 17)

DNA polymerase: an enzyme that carries out the synthesis of new DNA. (pg. 160)

dominant allele: the version of a gene that expresses its trait even in the company of a contrasting recessive allele. (pg. 150)

electronegativity: the measure of an atom's ability to attract electrons toward itself in a chemical bond. (pg. 41)

electron transport chain: a series of protein complexes that transfer electrons, generating an electrochemical gradient which is used to produce ATP. (pg. 185)

endocytosis: a type of active transport where cells use vesicles to move large amounts of substances into the cell. (pg. 119)

eukaryotic cell: cells that possess a distinct nucleus and various specialized organelles, such as mitochondria and the endoplasmic reticulum, each composed of specialized, membrane-bound compartments. (pg. 127)

exocytosis: a type of active transport is when cells use vesicles to move large amounts of substances out of the cell. (pg. 118)

exons: pieces of proteins that are spliced together during RNA processing. (pg. 166)

facilitated transport: transport that uses a protein to move across the membrane. (pg. 118)

final electron acceptor: the last molecule in an electron transport chain to receive electrons, crucial for

C

maintaining electron flow and ATP production. (pg. 185)

G1 phase: a period of interphase during which the cell undergoes significant growth marking the beginning of the cell cycle following cell division characterized by active production of proteins and other molecules necessary to carry out the cell's functions. (pg. 137)

G2 phase: a period of interphase during which the cell grows rapidly to prepare for cell division. (pg. 139)

gas phase: the state of matter where atoms or molecules are far apart and are not held together by intermolecular forces, resulting in a highly disordered and random structure. (pg. 31)

glycogen: a polysaccharide commonly found in animal tissues as a storage form of glucose. (pg. 86)

heterogeneous mixture: mixture in which the composition is not uniform throughout. (pg. 35)

homogenous mixture: mixture in which the composition is uniform throughout. (pg. 34)

hydrogen bonds: weak bonds that occur between polar molecules where the positive hydrogen end of one molecule attracts the negative end of molecules, causing them to stick together (pg. 48)

hydrolysis: involves breaking down a complex molecule into smaller units by adding water. (pg. 77)

hydrophobic: molecules that do not mix with water. (pg. 61)

independent assortment: describes how different genes independently separate from one another when reproductive cells develop. (pg. 149)

independent variable: the variable that the researcher intentionally manipulates or changes to observe its effect on the dependent variable. (pg. 17)

integral membrane proteins: permanently embedded in the phospholipid bilayer, these proteins play a major role in transporting materials across the membrane, attaching cells to one another, and acting as molecular markers for cellular communication. (pg. 116)

introns: pieces of mRNA that are removed during RNA processing. (pg. 166)

ion: an atom that doesn't have an equal number of protons and electrons and therefore has a charge. (pg. 27)

ionic bond: a bond where one or more electrons are transferred from one atom to another. (pg. 43)

isotope: an element where the number of protons has remained constant, but the number of neutrons is no longer equal to the number of protons. (pg. 26)

liquid phase: the state of matter where atoms or molecules are more loosely packed than in a solid but are still held together by intermolecular forces. (pg. 31)

meiosis: an essential cellular process that divides a diploid cell into four haploid cells, a step in sexual reproduction that contributes to genetic diversity. (pg. 146)

meniscus: the curved surface of water in a cylinder, caused by water creeping up slightly at the sides. (pg. 60)

mitosis: a fundamental process for cell division whose primary function is to enable the growth and repair of damaged cells in multicellular organisms by producing new cells that are genetically identical to the parent cell, ensuring the equal distribution of genetic material in two daughter cells. (pg. 143)

mitotic spindle: a microtubule-based structure that forms during cell division, orchestrating chromosome alignment and segregation into daughter cells. (pg. 143)

mixture: a type of matter that contains two or more substances that can be separated by physical means. (pg. 33)

molecule: the result of two or more atoms bonding together. (pg. 42)

mutation: any change in the sequence of DNA. (pg. 171)

nonpolar covalent bond: a bond occurring when valence electrons are shared equally between two atoms. (pg. 46)

nucleotides: the monomers of nucleic acids, including DNA and RNA. (pg. 155)

osmosis: the diffusion of water molecules from a high concentration to a low concentration. (pg. 121)

peripheral membrane proteins: proteins that are added or removed from the membrane relatively easily and are usually temporary, often used in the regulation of integral proteins, attachment to the cytoskeleton or extracellular matrix, communication through signal transduction, and as enzymes. (pg. 117)

pH: the "power of hydrogen" is a measure of the concentration of hydrogen ions (H+) in solution. (pg. 66)

phagocytosis: when a cell eats another cell; a type of endocytosis. (pg. 119)

phospholipids: are a type of lipid with a hydrophilic head and hydrophobic tails that form the basic structure of cell membranes. (pg. 91)

placebo: a substance or treatment that has no therapeutic effect but is given to a patient as if it were a real treatment. (pg. 17)

polar covalent bond: a bond occurring when the electrons are shared unequally between two atoms. (pg. 43)

polarity: uneven distribution of charges across a molecule, making one end positive (H) and the other negative (O). (pg. 59)

polar molecule: molecule that has electrically charged areas. (pg. 46)

prokaryotic cell: a simple, single-celled organism lacking a true nucleus and other membrane-bound organelles with genetic material freely floating in the cytoplasm. (pg. 127)

promoter: the location on DNA where transcription begins. (pg. 164)

pure substance: a type of matter that has a fixed composition and cannot be separated into simpler substances by physical means. (pg. 32)

recessive allele: the version of a gene that only emerges when it is paired with another recessive allele. (pg. 150)

ribosome: the cell organelle responsible for building proteins, primarily composed of a type of RNA called ribosomal RNA, or rRNA. (pg. 167)

RNA polymerase: the enzyme that carries out transcription in gene expression. (pg. 164)

segregation: part of the meiosis process where pairs of gene variants (alleles) are separated into different gametes, ensuring that each gamete carries only one allele for each gene. (pg. 149)

selectively permeable or semi-permeable: a process where some molecules can pass directly through the phospholipid bilayer, and others cannot, which allows cells to regulate which molecules are allowed to pass through. (pg. 117)

solid phase: the state of matter where atoms or molecules are tightly packed and held in a fixed position by intermolecular forces. (pg. 31)

solute: the substance in a solution that is dissolved in the solvent. (pg. 34)

solution: a special type of homogeneous mixture in which one substance (the solute) is dissolved in another substance (the solvent). (pg. 34)

solvent: the substance that does the dissolving in a solution. (pg. 34)

specific heat: the amount of heat needed to increase the temperature of 1 g of a substance by 1°C. (pg. 61)

S phase: part of the cell cycle where DNA replication occurs. (pg. 138)

starch: a colloquial term for polysaccharides, often referring to amylose. (pg. 86)

surface tension: a force that acts on the particles at the surface of a liquid. (pg. 60)

suspension: a type of heterogeneous mixture whose components are not evenly distributed throughout the mixture. (pg. 35)

tetrad: a structure formed during meiosis, consisting of two homologous chromosomes, each with two sister chromatids, totaling four chromatids that facilitates genetic recombination. (pg. 146)

translation: the process where ribosomes decode messenger RNA (mRNA) sequences into specific amino acid chains, forming proteins. (pg. 167)

transport proteins: proteins that move larger molecules, anything with a charge, or anything polar across the membrane. (pg. 117)

treatment group: a group that is exposed to the experimental treatment of the independent variable. (pg. 17)

valence electrons: the electrons in the outermost shell of an atom. (pg. 39)

C

Acknowledgements

Figure 1.1. "Biological Levels of Organization" by *General Biology* 2e. LibreTexts Biology.

Figure 1.2. "Eukaryote DNA" by Sponk, Tryphon, Magnus Manske, User: Dietzel65, LadyofHats (Mariana Ruiz), Radio89. Wikimedia Commons. (Wikimedia Commons, CC BY-SA 3.0).

Figure 1.5. "CO_2 Regulation Cycle" by Mr. Bothell at Meridian Technical Charter High School.

Figure 1.6. "The Scientific Method as an Ongoing Process" adapted from ArchonMagnus. (Wikipedia, CC BY-SA 4.0).

Figure 1.7. "How do vaccines work?" by the World Health Organization.

Figure 1.8. "Prevalence of Five Developmental Disabilities, Metropolitan Atlanta Developmental Disabilities Surveillance Program, 1991–2010.

Figure 1.9. "The Structure of an Atom" by OpenStax.org, licensed under Creative Commons Attribution License v4.0.

Figure 1.10. "The Periodic Table of Elements" by OpenStax *Chemistry* 2e, licensed under Creative Commons Attribution License v4.0.

Figure 1.11. "Visual Connection" Human Biology Copyright © by Janet Wang-Lee is licensed under a Creative Commons Attribution-NonCommercial-ShareAlike 4.0 International License.

Figure 1.12. "Bohr Diagrams" by Rice University, OpenStax *Biology* 2e, licensed under Creative Commons Attribution License v4.0.

Figure 1.13. "Most Common States or Phases of Matter" by Rice University, OpenStax *Chemistry* 2e, licensed under Creative Commons Attribution License v4.0.

Figure 1.14. "Matter Classification Flow Chart" by Rice University, OpenStax *Chemistry* 2e, licensed under Creative Commons Attribution License v4.0.

Figure 1.15. "Mixtures and Pure Substances" by John Trombley. (Wikipedia, CC BY-SA 4.0)

Figure 2.1. "Oxyen Bohr Model" by Ahazard. sciencewriter. (Wikimedia Commons, CC BY-SA 4.0)

Figure 2.2. "Periodic Table Organic Chemistry" by Naturwiki. (Wikimedia Commons, CC BY-SA 4.0)

Figure 2.3. "Covalent and Ionic Bonds" by Robert Lancashire, UWI. (Wikimedia Commons, CC BY-SA 3.0)

Figure 2.4. "The Formation of NaCl Through Ionic Bond" by Rice University, OpenStax *Biology* 2e, licensed under a Creative Commons Attribution 4.0 International License.

Figure 2.5. "Types of Covalent Bonds" by Robert Lancashire, UWI. (Wikimedia Commons, CC BY-SA 3.0)

Figure 2.6. "Covalent Bonding" by OpenStax *Anatomy and Physiology*. (Wikimedia Commons, CC BY-SA 3.0)

Figure 2.10. "Space Filling Model of the Combustion Reaction of Methane with Oxygen" by Rice University, OpenStax *Chemistry* 2e, licensed under Creative Commons Attribution License v4.0.

Figure 2.11. "Three Fundamental Chemical Reactions" by OpenStax *Anatomy and Physiology*. (Wikimedia Commons, CC BY-SA 3.0)

Figure 3.1. "Model Hydrogen Bonds in Water," translated by Michal Maňas (User:snek01). Wikimedia Commons.

Figure 3.2. "Surface Tension of Water," University of Hawai'i Anatomy & Physiology, based on OpenStax *Anatomy and Physiology*, which is licensed under a Creative Commons Attribution 4.0 International License.

Figure 3.3. "Capillary Attraction Repulsion" licensed with PD-ScottForesman.

Figure 3.4. "Water Molecules and Their Interaction with Salt" by Water Science School. USGS.

Figure 3.5. "pH Scale" by OpenStax College *Anatomy and Physiology*, Connexions Website. This file is licensed under the Creative Commons Attribution 3.0 Unported license.

Figure 3.14. "Medical anatomy illustration of human gastrointestinal digestive system including intestines or gut" by Christos Georghiou. Adobe Stock.

Figure 4.8. "Carbohydrates Digestion. Maltase, Sucrase, and Lactase enzymes catalyze disaccharides maltose, lactose, and sucrose to monosaccharides, glucose, galactose, and fructose molecules." Vector Illustration. By PH-HY. Adobe Stock.

Figure 4.10. “Synthesis of a Triglyceride” by Rice University, OpenStax, *Biology* 2e, licensed under Creative Commons Attribution License v4.0.
Figure 5.2. “Amino acids chart” by Biochemlife. Wikimedia Commons. (Wikimedia Commons, CC BY-SA 4.0)
Figure 5.5. “Protein structure. Amino acid chain molecule scheme. Primary structure” by inspiring.team. Adobe stock.
Figure 5.6. “Secondary Structure of Proteins” by CNX OpenStax. (Wikimedia Commons, CC BY-SA 4.0)
Figure 5.7. “Tertiary Structure of Proteins” by Rice University. OpenStax *Biology* 2e, licensed under Creative Commons Attribution License v4.0.
Figure 5.8. “Mechanism of enzyme action with substrate and product complexes. Enzyme Kinetics” by sakurra. Adobe Stock.
Figure 5.9. “The Effect of Enzymes on Activation Energy of a Reaction” by OpenLab at City Tech, OpenStax *Biology*, licensed under Creative Commons Attribution License v4.0.
Figure 5.10. “Scheme of enzyme action on a substrate” by tang90246. Adobe Stock.
Figure 5.11. “Denaturation and renaturation of Proteins” by Aldona. Adobe Stock.
Figure 5.12. “OSC Microbio 08 01 ApoHolo” by CNX OpenStax. (Wikimedia Commons, CC BY-SA 4.0)
Figure 5.13. “OSC Microbio 08 01 EnzInh” by CNX OpenStax. (Wikimedia Commons, CC BY-SA 4.0)
Figure 6.4. “Fluid Mosaic Model of the Cell Membrane” by OER Atlanta. Online Textbook for Biol 395 Copyright © 2024 by Jantina Toxopeus is licensed under a Creative Commons Attribution 4.0 International License.
Figure 6.5. “Open and closed conformations of ion channels” by Efazzari. (Wikimedia Commons, CC BY-SA 4.0)
Figure 6.6. “Scheme simple diffusion in cell membrane” by LadyofHats Mariana Ruiz Villarreal. Wikimedia Commons.
Figure 6.7. “Scheme facilitated diffusion in cell membrane” by LadyofHats Mariana Ruiz Villarreal. Wikimedia Commons.
Figure 6.8. “Phagocytosis and Exocytosis by Biostudent19. (Wikimedia Commons, CC BY-SA 4.0)
Figure 6.11. “Osmotic pressure on blood cells diagram” by LadyofHats. Wikimedia Commons.
Figure 7.1. “Eukaryotic vs Prokaryotic cells, educational biology illustration diagram, transparent background. Microbiology scheme with cell type examples. Cell membranes, cytoplasm, chromosomes” by VectorMine. Adobe Stock.
Figure 7.2. “Endomembrane system diagram” by Mariana Ruiz LadyofHats. Wikimedia Commons.
Figure 7.3. “The three types of cytoskeletal filaments” by Fundamentals of Cell Biology. Manifold Scholarship.
Figure 7.4. “Eukaryotic vs Prokaryotic cells, educational biology illustration diagram, transparent background. Microbiology scheme with cell type examples. Cell membranes, cytoplasm, chromosomes” by VectorMine. Adobe Stock.
Figure 7.6. “HeLa cells stained with Hoechst” by TenOfAllTrades at English Wikipedia. Wikimedia Commons.
Figure 7.7. “Cell Cycle” by Zephyris. (Wikimedia Commons, CC BY-SA 3.0)
Figure 7.8. “Chromosome Terminology” by Christinelmiller. (Wikimedia Commons, CC BY-SA 4.0)
Figure 7.9. “Cell Cycle Checkpoints” by Rice University, OpenStax *Concepts of Biology*, is licensed under Creative Commons Attribution License v4.0.
Figuro 8.1. “Chromosome Terminology” by Christinelmiller. (Wikimedia Commons, CC BY-SA 4.0)
Figure 8.2. “Mitosis cells sequence” by LadyofHats. (Wikimedia Commons, CC BY-SA 1.0)
Figure 8.3 “Overview of the Cell Cycle Checkpoints and Phases” by Rice University, OpenStax *Concepts of Biology*, licensed under Creative Commons Attribution License v4.0.
Figure 8.4. “Meiosis Stages” by Ali Zifan. (Wikimedia Commons, CC BY-SA 4.0)
Figure 8.5. “Meiosis Stages” by Ali Zifan. (Wikimedia Commons, CC BY-SA 4.0)
Figure 8.6. “Biological Life Cycle of Humans” by Aiden cormier. (Wikimedia Commons, CC BY-SA 4.0)
Figure 9.1. “Nucleotide” by DrTOsborne. (Wikimedia Commons, CC BY-SA 3.0)
Figure 9.2. “DNA infographics” by vector punch. Adobe Stock.
Figure 9.3. “DNA transcription” by National Human Genome Research Institute.
Figure 9.4. “Semi-conservative model of DNA Replication” by Rice University, OpenStax *Concepts of Biology*, licensed under Creative Commons Attribution License v4.0.
Figure 9.5. “DNA Polymerase DNA Replication” by Christinelmiller. (Wikimedia Commons, CC BY-SA 340)
Figure 9.6. “Central Dogma of Molecular Biochemistry with Enzymes” by Dhorspool (talk) Daniel Horspool. (Wikimedia Commons, CC BY-SA 3.0)
Figure 9.7. “Transcription Bubble” by MSPaintDrawings. (Wikimedia Commons, CC BY-SA 4.0)

Figure 9.8. "Splicing" by OpenStax. (Wikimedia Commons, CC BY-SA 4.0)
Figure 9.9. "Proteinsynthesis" by Mayera. (Wikimedia Commons, CC BY-SA 3.0)
Figure 9.10. "Types of RNA" by Christinelmiller. (Wikimedia Commons, CC BY-SA 4.0)
Figure 9.11. "Ribosome mRNA translation" by LadyofHats. Wikimedia Commons.
Figure 9.12. "Amino Acid Codon Table" by Scott Henry Maxwell. (Wikimedia Commons, CC BY-SA 4.0)
Figure 9.13. "Silent Mutation" by Rice University, OpenStax *Microbiology*, licensed under Creative Commons Attribution License v4.0.
Figure 9.14. "Substitution Mutation" by Rice University, OpenStax *Microbiology*, licensed under Creative Commons Attribution License v4.0.
Figure 9.15. "Frame Shift Mutation" by Rice University, OpenStax *Microbiology*, licensed under Creative Commons Attribution License v4.0.
Figure 10.4. "ATP cycle," by Eunice Laurent. (Wikimedia Commons, CC BY-SA 4.0)
Figure 10.5. "Vector diagram of Mitochondria. Cross-section view. Medical infographics" by L.Darin. Adobe Stock.
Figure 10.6. "Electron Carrier Nicotinamide Adenine Dinucleotide (NAD) + Hydrogen (H), NADH" by Rice University, OpenStax *Biology* 2e, licensed under Creative Commons Attribution License v4.0.
Figure 10.7. "FAD FADH2 equilibrium" by DMacks. (Wikimedia Commons, CC BY-SA 3.0)
Figure 10.11. "Cellular respiration" by Zlir'a. (Wikimedia Commons, CC BY-SA 4.0)
Figure 10.13. "Pyruvate Oxidation" by British Columbia/Yukon Pressbooks. Pressbooks Guide Copyright © 2016–2022 by OER Production Team is licensed under a Creative Commons Attribution 4.0 International License.
Figure 10.14. "The Citric Acid Cycle" by Rice University, OpenStax *Concepts of Biology*, licensed under Creative Commons Attribution License v4.0.
Figure 10.15. "The Electron Transport Chain, Chemiosmosis, and ATP Synthase" by Rice University, OpenStax *Biology* 2e, licensed under Creative Commons Attribution License v4.0.
Figure 10.16. "ATP Synthase" by Rice University, OpenStax *Biology* 2e, licensed under Creative Commons Attribution License v4.0.
AppA.1. "Periodic Table of Elements" by National Center for Biotechnology Information (2025). National Library of Medicine.
AppB.1. "Common Organic Functional Groups"

www.ingramcontent.com/pod-product-compliance
Lightning Source LLC
Chambersburg PA
CBHW080032210826
50197CB00009B/450
* 9 7 8 1 9 5 5 4 9 9 5 7 6 *